MEASURE
AND INTEGRATION

MEASURE AND INTEGRATION

A Concise Introduction
to Real Analysis

Leonard F. Richardson

A JOHN WILEY & SONS, INC., PUBLICATION

Published by John Wiley & Sons, Inc., Hoboken, New Jersey.
Published simultaneously in Canada.

For general information on our other products and services or for technical support, please contact our Customer Care Department within the United States at (800) 762-2974, outside the United States at (317) 572-3993 or fax (317) 572-4002.

Wiley also publishes its books in a variety of electronic formats. Some content that appears in print may not be available in electronic format. For information about Wiley products, visit our web site at www.wiley.com.

Library of Congress Cataloging-in-Publication Data:

Richardson, Leonard F.
  Measure and integration : a concise introduction to real analysis / Leonard F. Richardson.
    p. cm.
  Includes bibliographical references.
  ISBN 978-0-470-25954-2 (cloth)
  1. Lebesgue integral. 2. Measure theory. 3. Mathematical analysis. I. Title.
  QA312.R45 2008
  515'.42—dc22   2009009714

Printed in the United States of America.

10 9 8 7 6 5 4 3 2 1

To Joan, Daniel, and Joseph

CONTENTS

PREFACE

The purpose of this textbook is to provide a concise introduction to *Measure and Integration*, in the context of abstract measure spaces. It is written in the hope that it has sufficient versatility to provide the prerequisites for beginning graduate courses in harmonic analysis, probability, and functional analysis. For both mathematical and pedagogical reasons, it is helpful to the student to emphasize examples and exercises from the real line and Euclidean space. This emphasis also reflects the special theorems that should be established for the cases of the line and Euclidean space.

The author's department spans diverse branches of both pure and applied mathematics, making it desirable to teach sufficient measure and integration in one semester. This helps to make room in the introductory part of the graduate curriculum for both complex analysis and functional analysis. The choice of textbook for achieving this goal, covering both abstract measure spaces and Euclidean space in one semester, has been problematic.

In 1963, the author was fortunate to be a student in the course "Measure and Integration" taught by Professor Shizuo Kakutani (1911–2004) at Yale University. His course achieved all the objectives cited above. It also succeeded in conveying to the student a sense of awe for the profound insights the subject has to offer for mathematics.

The author had hoped that a book on this subject might be written by Professor Kakutani himself. Since this did not happen, and because the author's notes from Professor Kakutani's course seemed potentially useful if converted into a textbook, this book has been written. Many exercises have been added, these being very important to a sound core graduate course in analysis.

Several favorite topics in real analysis augment the original course. There are two concluding chapters, intended for the reader who would like to continue this study after the end of the semester. These chapters provide some favorite applications to functional analysis, and to the study of translation-invariant function spaces on the line and the circle. The excursion in the latter direction concludes with a brief proof that $L^2(\mathbb{R})$ is irreducible under the combined actions of translation in the (real) domain, and rotation in the (complex) range, of the functions. There is a very brief introduction to the Heisenberg Commutation Relation and the role of the Heisenberg group in the Stone-von Neumann theorem, which is not proven here. The role of translation invariance in real analysis, beginning with Lebesgue measure and the Lebesgue integral themselves, reflect the author's interest in the underlying group-theoretic structures in analysis.

ACKNOWLEDGMENTS

Real analysis, and how to teach it, is a perennial topic of discussion among analysis professors, and I have benefited in more ways than I could enumerate from conversations with colleagues for 35 years. I feel that I have benefited in the writing of this book especially from discussions with Professors Jacek Cygan, Mark Davidson, Raymond Fabec, and P. Sundar.

It has been a great pleasure and privilege to teach measure and integration to numerous graduate students over the years. They have all contributed to my enjoyment of the subject and to the choices that went into this book. For the autumn semesters of 2007 and 2008, incoming graduate classes at LSU learned measure and integration from a preliminary version of this book. These fine students kindly pointed out numerous typos, and they asked questions that enabled me to improve the exposition. Some found interesting resources in the digital literature. Especially helpful to me were two advanced graduate assistants, Jens Christensen and Anna Zemlyanova. They did a superlative job of grading the weekly homework assignments collected during these two terms, and their observations were especially helpful to me in improving the text.

This book began with my own notebook from Professor Shizuo Kakutani's course in measure theory, taught in 1963 at Yale University. It was Professor Kakutani's way to leave it to students to write detailed notes for the course, filling in many proofs that were omitted in class. This work was the principal assignment whereby students earned academic credit. I have added numerous exercises to the original course, believing these to be essential to the teaching of present-day graduate students. Some exercises require the student to complete a proof or to extend a theorem, thereby incorporating in a formal manner duties of the kind that Professor Kakutani left as an assignment. Many other exercises are problems typical of PhD qualifying examination questions in real analysis. These are intended show students how the subject of measure and integration is used in analysis, and to help them prepare for milestone tests such as qualifiers.

No doubt the best features of this text were learned from Professor Kakutani. Also, there are ideas borrowed from several favorite books, which are cited in the References. These are recommended highly for further reading. The supplementary topics in the last chapter, and a number of the exercises in the text, reflect my interest in the role of group actions in analysis, which in this course means mainly the action of the real line or the Euclidean vector group. In this I have benefited from the teaching my own major professor at Yale from 1967 to 1970, Professor George D. Mostow, to whom I am indebted beyond words. The shortcomings and errors of this book are, of course, exclusively my own. I welcome gratefully corrections and suggestions from readers.

I am grateful to John Wiley & Sons for the opportunity to offer this book, as well as the course it represents and advocates, to a wider audience. I appreciate especially the role of Ms. Susanne Steitz-Filler, the Mathematics and Statistics Editor of John Wiley & Sons, in making this opportunity available. She and her colleagues provided valued advice, support, and technical assistance, all of which were needed to transform a professor's course notes into a book.

L.F.R.

INTRODUCTION

The reader of this book is likely to be a first-semester graduate student. The author advises this reader, gently but firmly, to expect that what follows is an intense and sustained exercise in logical reasoning. The effort will be great, but the reader may find that the reward is greater and that this subject is an important cornerstone of modern mathematics.

Strategies employed commonly by undergraduates for the study of mathematics may not be appropriate for the graduate student. Because the reader is assumed to be a beginning graduate student in mathematics, I offer here some advice regarding how to study mathematics in graduate school.

In reading the proofs of theorems in this text or in the study of proofs presented by one's teacher in class, the student must understand that what is written is much more than a body of facts to be remembered and reproduced upon demand. Each proof has a story that guided the author in its writing. There is a beginning (the hypotheses), a challenge (the objective to be achieved), and a plan that might, with hard work, skill, and good fortune, lead to the desired conclusion. Thus each step of a proof should be seen by the student as meaningful and purposeful, and not as a procedure that is being followed blindly. It will take time and concerted effort for the student to learn to think about the statements and proofs of the presented theorems in this light. Such practice will cultivate the ability to do the exercises as well in a fruitful manner. With experience in recognizing the story of the proof or problem at hand, the student will be in a position to develop his or her own technique through the work done in the exercises.

The first step, before attempting to read a proof, is to read the statement of the theorem carefully, trying to get an overall picture of its content. The student should make sure that he or she knows precisely the definition of each term used in the statement of the theorem. Without that information, it is impossible to understand even the claim of the theorem, let alone its proof. If a term or a symbol in the statement of a theorem or exercise is not recognized, look in the index! Write down what you find.

After clarifying explicitly the meaning of each term used, if the student does not see what the theorem is attempting to achieve, it is often helpful to write down a few examples to see what difficulties might arise, leading to the need for the theorem. *Working with examples is the mathematical equivalent of laboratory work for a natural scientist.* Indeed, the student should accumulate a tool kit of examples that illustrate what may go wrong with a theorem if the hypotheses are not satisfied. One also should learn examples that show why stronger conclusions than those stated may fail to be valid with the given hypotheses. Many exercises in the text are intended to provide examples that assist one in these ways to grasp the meaning of the theorems, and the student must understand that the exercises are a very important component of the study of measure and integration. The experience gained from the exercises will help the student to understand which steps in the solution of a problem require careful justification.

CHAPTER 1

HISTORY OF THE SUBJECT

1.1 HISTORY OF THE IDEA

In a broad sense, much of mathematics is devoted to the decomposition, or analysis, of a whole entity into its component parts and the reconstruction, or synthesis, of the whole from its parts. There is in this an expectation that the whole is somehow equal to the sum of its parts, which are simpler individually. We discuss briefly a few examples of this aspect of mathematics.

It was nearly three thousand years ago that Babylonian astronomers successfully predicted the times of lunar and solar eclipses by expressing these complicated events as summations of numerous simpler periodic events. The predictions were fairly accurate, to the extent of predicting eclipses that would be visible at least from some part of the world. That remarkable achievement may be interpreted as the first appearance on Earth of harmonic (or Fourier) analysis.

Measurement has been a special interest for mathematicians and scientists since the early days of civilization. Two thousand years ago, the classical geometers of Greece made profound contributions to the study of measurement. A line segment could be *measured by a shorter segment* if the short one could be laid off end to end in such a way that the long segment would be seen to be an exact positive integer

Measure and Integration: A Concise Introduction to Real Analysis. By Leonard F. Richardson

multiple of the short one. Two segments were called *commensurable* if both could be measured by a *common* shorter segment, or *unit* of length. And the discovery by Pythagoras or his associates of *incommensurable* segments was a momentous event in the history of mathematics. Inherent in the definition Greek geometers used for measurement of segments was the concept that the measure of the *whole segment must be the sum of the measures of its nonoverlapping parts*, as indicated by the lengths marked off according to the shorter segment.

The Greek geometers pushed this technique much farther, successfully studying the circumference and area of a circle and the surface area and volume of a sphere. These challenges were met by the application of what is called the *principle of exhaustion*. The method was to approximate a geometrical measurement from below, and at each successive iteration of the approximation technique, at least of half of what remained to be counted was to be included. (Greek geometers understood that if too little were taken at each stage, even an endless succession of steps might not approach the goal.) For example, the area of a circle is approximated from within by means of an inscribed 2^n-sided regular polygon. As n increases, the area of the resulting 2^n-sided polygon grows in a computable manner. (See Figure 1.1.) We begin with an inscribed square consuming the bulk of the area of the circle. The next polygon is an octagon, which adds the areas of four thin triangles to that of the square. At each stage, one can construct the 2^{n+1}-gon by bisecting the arcs that are

Figure 1.1 Area of a circle by exhaustion.

subtended by the sides of the inscribed 2^n-gon. In effect, the $(n + 1)$th stage of the approximation results from adding a small increment to what was obtained at the nth stage. Such classical achievements in geometry may have been the first instances in which a measure (of area, for example) was sought that would be *countably additive*, meaning that the measure of the whole *should be* the sum of the measures of its *infinite sequence* of nonoverlapping parts.

Greek geometers and philosophers may not have been entirely convinced of the validity of the principle of exhaustion. The expression of some dissatisfaction with this geometric technique may have been the purpose of the famous paradox of Zeno, in which the legendary warrior Achilles was pitted unsuccessfully in a footrace with a persistent tortoise.

The notion that the whole should be the sum of even an infinite sequence of its nonoverlapping parts appears very strongly in modern analysis, both pure and applied. For example, in 1822 Joseph Fourier presented a seminal paper on the heat equation to the French Academy, in which he introduced the use of infinite trigonometric series for the purpose of determining the solution of the heat equation on a finite interval, or rod [6]. Fourier's method required that the hypothetical solution function be *expressible* as the sum of an infinite trigonometric series:

$$f(x) = \sum_{0}^{\infty}(a_n \cos nx + b_n \sin nx).$$

Unfortunately, these so-called Fourier series can diverge for very large sets of numbers x in the domain of even a rather nice function f. Yet, if one ignored the embarrassing reality that $f(x)$ *need not be the sum of its parts*, Fourier's method actually worked. And the search was on for suitable concepts and tools to analyze correctly the right classes of functions for which the Fourier series would converge in a useful sense to the functions being represented.

The efforts took place on a grand scale. Even the theory of sets was invented by Georg Cantor for the purpose of analyzing sets of convergence and divergence of Fourier series. Though Cantor's approach was not sufficient for the needs of Fourier series, it became a cornerstone of modern mathematics.

Early in the twentieth century Henri Lebesgue invented his new and very refined concept of the integral, based on the *measure* of suitable subsets of the line. Lebesgue measure became the foundation not only for Fourier analysis, but also for probability, and for functional analysis which permeates modern analysis.

1.2 DEFICIENCIES OF THE RIEMANN INTEGRAL

The Riemann integral is the integral of elementary calculus. It is the integral developed intuitively by Newton and Leibnitz and put to great use in the classical sciences. Before undertaking the considerable work of developing the Lebesgue integral, the reader and student need to become acquainted with the deficiencies of the Riemann integral. This will motivate the effort that follows.

First, we review the definition of the Riemann integral of a bounded real-valued function f on a closed, finite interval $[a, b]$ of the real line.

Definition 1.2.1 A partition P is an ordered list of finitely many points starting with a and ending with $b > a$. Thus $P = \{x_0, x_1, \ldots, x_n\}$, where

$$a = x_0 < x_1 < \cdots < x_n = b.$$

These points are regarded as partitioning $[a, b]$ into n contiguous subintervals, $[x_{i-1}, x_i]$, $i = 1, \ldots, n$. The length of the ith subinterval is given by $\Delta x_i = x_i - x_{i-1}$. The *mesh* of the partition is denoted and defined by

$$\|P\| = \max \{\Delta x_i \mid i = 1, 2, \ldots n\}.$$

Definition 1.2.2 Let f be *any* bounded function on $[a, b]$ and let P be any partition of $[a, b]$. Let

$$M_i = \sup \{f(x) \mid x \in [x_{i-1}, x_i]\} \text{ and } m_i = \inf\{f(x) \mid x \in [x_{i-1}, x_i]\}.$$

Define the upper sum,

$$U(f, P) = \sum_{i=1}^{n} M_i \Delta x_i,$$

and the lower sum,

$$L(f, P) = \sum_{i=1}^{n} m_i \Delta x_i.$$

We say that f is Riemann integrable on $[a, b]$ with $\int_a^b f(x) \, dx = L$ if and only if *both* $L(f, P) \to L$ and $U(f, P) \to L$ as $\|P\| \to 0$.

Note that M_i and m_i are real numbers in Definition 1.2.2 because f is bounded.

■ EXAMPLE 1.1

Since the set $\mathbb{Q}$ of rational numbers is countably infinite, the same is true of the set S of all rational numbers in $[a, b]$ for any $a < b$. So, write $S = \{q_n \mid n \in \mathbb{N}\}$. Now define the functions

$$f_n(x) = \begin{cases} 1 & \text{if } x \in \{q_1, \ldots, q_n\}, \\ 0 & \text{if } x \in [a, b]\backslash\{q_1, \ldots, q_n\}. \end{cases}$$

It is known from advanced calculus[1] that each function f_n lies in $\mathfrak{R}[a, b]$, the set of Riemann integrable functions on $[a, b]$. In the following exercises, the reader will prove that the pointwise limit of the sequence f_n is not Riemann integrable.

---

[1] See [20], for example.

EXERCISES

The exercises below refer to the functions f_n in Example 1.1.

1.1   Prove that each function f_n lies in $\mathfrak{R}[a, b]$, the set of Riemann integrable functions on $[a, b]$.

1.2   Prove that for each x in $[a, b]$, $f_n(x) \to 1_S(x)$, where

$$1_S(x) = \begin{cases} 1 & \text{if } x \in S, \\ 0 & \text{if } x \in [a, b] \backslash S, \end{cases}$$

the *indicator function* of the set S of rational numbers in $[a, b]$.

1.3   Prove that the function 1_S is not Riemann integrable. That is,

$$\lim_{n \to \infty} \int_a^b f_n(x)\, dx \neq \int_a^b \lim_{n \to \infty} f_n(x)\, dx,$$

because the latter integral does not exist.

The failure of the pointwise limit of a sequence of Riemann integrable functions to be Riemann integrable is considered a serious shortcoming of the Riemann integral. The following example will illustrate a deficiency that is shared by the Riemann integral and the Lebesgue integral that we will define.

■ **EXAMPLE 1.2**

Let

$$f_n(x) = \begin{cases} n & \text{if } 0 < x \leq \frac{1}{n}, \\ 0 & \text{if } \frac{1}{n} < x \leq 1, \\ 0 & \text{if } x = 0 \end{cases}$$

for all $n \in \mathbb{N}$. The reader should do the following exercise.

EXERCISE

1.4   Let f_n be as in Example 1.2. Prove that $f_n(x) \to f(x) \equiv 0$ pointwise on $[0, 1]$.

Also, it is clear that $f_n \in \mathfrak{R}[0, 1]$ for all n, and $f \in \mathfrak{R}[0, 1]$ as well. Yet

$$\int_0^1 f_n(x)\, dx \equiv 1 \to 1 \neq 0 = \int_0^1 f(x)\, dx.$$

Thus it occurs for some convergent sequences of functions that

$$\lim_{n \to \infty} \int_a^b f_n(x)\, dx \neq \int_a^b \lim_{n \to \infty} f_n(x)\, dx \tag{1.1}$$

even when all the integrals exist. For the Lebesgue integral, however, Theorem 5.3.1 will identify useful conditions under which equality would be guaranteed in Equation (1.1).

■ **EXAMPLE 1.3**

Let

$$f_n(x) = \begin{cases} \frac{1}{\sqrt{x}} & \text{if } \frac{1}{n} \leqslant x \leqslant 1, \\ 0 & \text{if } 0 \leqslant x < \frac{1}{n} \end{cases}$$

for each $n \in \mathbb{N}$. The reader should check that each f_n is Riemann integrable but that if $f(x) = \lim_{n \to \infty} f_n(x)$, then $f \notin \mathcal{R}[0,1]$ because f is not bounded. The reader should recall from elementary calculus that f is, however, *improperly* Riemann integrable. In Exercise 5.44 the reader will see a generalization of this example that satisfies Lebesgue convergence theorems but that cannot be corrected with improper Riemann integration.

1.3 MOTIVATION FOR THE LEBESGUE INTEGRAL

The Lebesgue integral begins with a seemingly simple reversal of the intuitively appealing process of Definition 1.2.2. Instead of partitioning the interval $[a, b]$ on the x-axis into subintervals and considering the range of values of a bounded function f on each small subinterval, Lebesgue began with the interval $[m, M]$ on the y-axis, where

$$M = \sup\{f(x) \mid x \in [a,b]\}, \quad \text{and} \quad m = \inf\{f(x) \mid x \in [a,b]\}.$$

Thus $P = \{y_0, y_1, \ldots, y_n\}$, where $m = y_0 < y_1 < \cdots < y_n = M$. Next, instead of forming a sum of the lengths Δx_i of the x-intervals weighted by the heights M_i or m_i, Lebesgue sought to form a sum of the heights, y_i, each weighted by some suitable concept of the length, or *measure* μ, of the set $f^{-1}([y_{i-1}, y_i])$, the set of points x for which $f(x) \in [y_{i-1}, y_i]$. The difficulty is that the set $f^{-1}([y_{i-1}, y_i])$ does not need to be an interval. Indeed, $f^{-1}([y_{i-1}, y_i])$ can be a very complicated subset of the x-axis.[2]

EXERCISE

1.5   Give an example of a real-valued function $f : \mathbb{R} \to \mathbb{R}$ for which

$$f^{-1}\left(\left[-\frac{1}{2}, \frac{1}{2}\right]\right) = \mathbb{R} \setminus \mathbb{Q},$$

the set of *irrational* numbers.

It turns out that the definition on the real line of the Lebesgue integral—a wonderful improvement upon the Riemann integral—is very simple once one has defined a

---

[2] The comparison of Riemann with Lebesgue integration has been likened to a story about a smart merchant who sorts money into denominations before counting the day's receipts. Riemann adds the figures as they come in, but Lebesgue sorts first according to values. Lebesgue integration is subtler, however, than this analogy suggests, because the sets $f^{-1}([y_{i-1}, y_i])$ can be very intricate indeed.

suitable concept of the positive real-valued measure of a subset of the line. The desired measure should agree with the concept of *length* when applied to a subset that is an interval. The key property that one needs for a concept of the measure of a set is that if one takes any infinite sequence of *mutually disjoint* sets E_i, one needs to have

$$\mu\left(\bigcup_{i=1}^{\infty} E_i\right) = \sum_{i=1}^{\infty} \mu(E_i).$$

That is, one needs a *countably additive measure* on subsets of the line which generalizes the concept of length of an interval. Unfortunately, no measure exists that agrees with the concept of the length of an interval and that can be defined on *all* the subsets of the line. Thus it turns out that defining the family of *Lebesgue measurable sets* is a very serious undertaking in the construction of the Lebesgue integral. And that is why we will begin our task in the next chapter with the definition of Lebesgue measurable sets and the definition of Lebesgue measure on those sets. This will turn out to be a *lengthy* task. (Confession: the pun *is* intended.)

We can see in advance how the Lebesgue integral will resolve some of the deficiencies of the Riemann integral. Suppose that we have defined already a countably additive Lebesgue measure that generalizes the concept of the length of an interval in the real line. The set $S = \mathbb{Q} \cap [a, b]$ of Exercise 1.3 is a countably infinite set. That is, the points of S can be arranged into a single infinite sequence: $S = \{s_n \mid n \in \mathbb{N}\}$. Each point is an interval of length zero. Thus it will need to be the case that the Lebesgue measure

$$l(S) = \sum_{n=1}^{\infty} l\{s_n\} = 0.$$

If the reader finds it believable in advance that the Lebesgue integral of a constant function on a measurable set will be that constant times the Lebesgue measure of the set, then

$$\int_a^b 1_S(x)\, dx = 1 \cdot 0 = 0$$

in the sense of Lebesgue integration.

The reader can understand at this point why Lebesgue measure is required to be *only* countably additive. If Lebesgue measure were to be *uncountably* additive, [3] then every set would have measure zero because every set is a disjoint union of singleton sets. Thus the theory of Lebesgue measure would collapse.

It will be seen in the coming chapters that the Lebesgue measure of each interval on the line will be its Euclidean length, that each Riemann integrable function will still be Lebesgue integrable, and that the value of that integral will be unchanged. Thus the reader is advised *not* to forget everything that he or she has learned before!

---

[3] One could define a concept of the sum of an uncountable family $\{x_a \mid a \in A\}$ of nonnegative real numbers indexed by an uncountable set A. For example, the sum could be taken to mean the supremum of the sums over all countable subsets of A. It is a simple exercise to show that, with this definition, a sum must be infinite unless $x_a = 0$ for all a outside some countable subset of A.

■ **EXAMPLE 1.4**

The fact that the Lebesgue measure of a countable set, such as the set of rational numbers, is zero will resolve another shortcoming of the Riemann integral. Let $S = \mathbb{Q} \cap [0, 1] = \{q_n \mid n \in \mathbb{N}\}$, as before. Define the functions

$$f_n(x) = \begin{cases} n & \text{if } x \in \{q_1, \ldots, q_n\}, \\ 0 & \text{if } x \in [0, 1]\backslash\{q_1, \ldots, q_n\}. \end{cases}$$

It is easy to see that $f_n(x)$ diverges to ∞ on the dense set S, whereas $f_n(x) \to 0$ at every other value of $x \in [0, 1]$. We can define a function

$$f(x) = \begin{cases} \infty & \text{if } x \in S, \\ 0 & \text{if } x \in [0, 1]\backslash S. \end{cases}$$

This function f is not real-valued at the points of S—we say that it is *extended* real-valued. But because the set S has Lebesgue measure zero, it will turn out that f is Lebesgue integrable and that $\int_0^1 f(x)\, dx = 0$, in the sense of Lebesgue. Thus we do have

$$\int_0^1 \lim_n f_n(x)\, dx = \lim_n \int_0^1 f_n(x)\, dx,$$

despite the fact that the pointwise limit of f_n exists only in the *extended* real-valued sense. Here we have benefited from the fact that the functions f_n are *uniformly bounded*, except on a set of measure zero. The reader should note that the function f is not even *improperly* Riemann integrable in any plausible sense.

Before proceeding to the task at hand, we explain why it is necessary to develop both Lebesgue measure and the Lebesgue integral for functions mapping a domain X that is an abstract set into the set $\mathbb{R}$ of real numbers. One reason for working at this level of generality, in which X is simply a set (not necessarily a set of real numbers) and $f : X \to \mathbb{R}$, is that it is important to define the Lebesgue integral for functions of several variables. That is, we wish also to be able to integrate $f : \mathbb{R}^n \to \mathbb{R}$. An element of $\mathbb{R}^n$ is not a real number, but rather an n-tuple of real numbers. Moreover, in higher analysis, both pure and applied, it is necessary to work with functions defined on groups, such as the important classical groups of matrices, and this requires knowledge of Lebesgue integration on abstract sets. Moreover, the study of Fourier inversion for functions defined on groups requires the introduction of a measure on what is called the *dual object*[4] of the group, and the Fourier transform must be integrated on that object.

---

[4]The dual object of a group is the set of all unitary equivalence classes of irreducible unitary representations of the underlying group. When endowed with a usable topology, such objects can be quite complex topologically.

Finally, there is a very important motivation for the abstract study of measure theory from probability. In a *probability model*, the outcomes of an experiment are pictured as points in a so-called *sample space* X. An *event* is conceptualized as a *subset* $E \subseteq X$. The idea behind this is that E denotes the event that the experiment yields a result that is an element of E. For example, X could be the real line. The experiment could be measuring the temperature of the mathematics classroom at 3 P.M. on a certain day. The interval $E = [80, 90]$ would represent the event that the temperature turns out to be between $80°$ and $90°$F.

In probability theory, one wishes very strongly to have a concept of the probability $\mu(E)$ that has the following properties:

- The probability $\mu(E) \in [0, 1]$ for each event E.

- The probability measure μ is additive on all countably infinite sequences of mutually disjoint events.

As discussed above, this cannot be done for all subsets of $\mathbb{R}$—at least not with a measure that generalizes reasonably the length of an interval. Thus for probability also, we must study the concept of measurable sets and the measure of such a set.

The sample space for a probability model need not be a subset of the real line. For example, Brownian motion is the type of motion exhibited by a particle suspended in a fluid. In order to study Brownian motion by means of probability theory, one must place a measure on the set of all possible *paths* that a Brownian motion may follow. [5]

Since the sample space of an experiment could be a set quite different from $\mathbb{R}$, we must develop the theory of measure and integration on abstract sets.

---

[5] It turns out that these paths are continuous, but they are nowhere differentiable with probability 1!

CHAPTER 2

FIELDS, BOREL FIELDS, AND MEASURES

In this chapter we introduce the families of subsets of a general set X on which we will be able to ascertain the existence of a countably additive measure. The concept will agree with that of the length of an interval in the real line, and with that of the volume of a rectangular box in Euclidean space of arbitrary finite dimension.

2.1 FIELDS, MONOTONE CLASSES, AND BOREL FIELDS

Let X be an abstract set. We will denote by $\mathfrak{P}(X)$ the *power set* of X, which is the set of all the subsets of X.

Definition 2.1.1 A nonempty family $\mathfrak{A} \subset \mathfrak{P}(X)$ is called a *field* of sets (also called an *algebra* of sets) provided that for all $A, B \in \mathfrak{A}$ we have

$$A \cup B \in \mathfrak{A},$$
$$A \cap B \in \mathfrak{A},$$
$$X \backslash A \in \mathfrak{A}.$$

Measure and Integration: A Concise Introduction to Real Analysis. By Leonard F. Richardson
Copyright © 2009 John Wiley & Sons, Inc.

Often, we will begin with a family of sets that we wish to have as elements of a field of sets, and we would like to know the *minimal* field that contains these sets.

Definition 2.1.2 Let $\mathfrak{A}$ be any family of subsets of X. Define $\mathbb{F}(\mathfrak{A})$, the *field generated* by $\mathfrak{A}$, to be the intersection of all the fields that contain $\mathfrak{A}$.

To see that there exists a field containing $\mathfrak{A}$, do Exercise 2.6. One can see that $\mathbb{F}(\mathfrak{A})$ is a field, since if A and B are in every field that contains $\mathfrak{A}$, the same is true of $A \cup B, A \cap B, X\backslash A$, and $X\backslash B$.

Observe that $\mathbb{F}(\mathfrak{A})$ is contained in every field that contains $\mathfrak{A}$. Thus it is reasonable to call $\mathbb{F}(\mathfrak{A})$ the *minimal* field that contains $\mathfrak{A}$.

■ **EXAMPLE 2.1**

Let $X = [0, 1) = \{x \mid 0 \leqslant x < 1\}$ be the real unit interval that is left-closed and right-open. Let $\mathfrak{E}$ denote the family of all the disjoint unions of *finitely* many intervals of the form $[a, b) \subseteq [0, 1)$.[6] It is easy to see that $\mathfrak{E}$ is a field since both the union and the complement of two intervals of the given form will always have the same form. This field will play an important role in the development of Lebesgue measure on the real line, and it is called the field of *elementary sets* in $[0, 1)$.

The reader should check, as an example, that the field generated by the set of all intervals of the form $[a, b) \subseteq [0, 1)$ is the field $\mathfrak{E}$ of all elementary sets in $[0, 1)$.

It is easy to generalize this example to define the field of elementary sets in any finite interval $[-N, N)$, for $N \in \mathbb{N}$. We can take an elementary set to be any finite union of left-closed, right-open intervals in $[-N, N)$.

EXERCISES

2.1   Show that the concept of a field of sets would be the same if we had omitted closure under unions from the three criteria listed in Definition 2.1.1.

2.2   Show that if $\mathfrak{A}$ is a field of subsets of X, then $X \in \mathfrak{A}$ and the empty set $\varnothing \in \mathfrak{A}$.

It is easy to see that a field is a family $\mathfrak{A}$ of subsets of X that is closed under the operations of taking unions and intersections of finitely many members of $\mathfrak{A}$ as well as closed under complementation.

Definition 2.1.3 A family $\mathfrak{A} \subseteq \mathfrak{P}(X)$ is called a *monotone class* if and only if it has the following two properties:

    i. If $A_1 \subseteq A_2 \subseteq \ldots \subseteq A_n \subseteq \ldots$ is an *increasing sequence* of sets $A_n \in \mathfrak{A}$, then $\bigcup_{n=1}^{\infty} A_n \in \mathfrak{A}$.

---

[6]Left-closed, right-open intervals are used only to make it easy to describe a *field* of elementary sets. The convenience is that the complement of $[a, b) \subseteq [0, 1)$ is a union of finitely many intervals of the same form. Later we will define the Borel sets and we will see that all intervals, and all open sets and closed sets, are among the Borel sets.

ii. If $A_1 \supseteq A_2 \supseteq \ldots \supseteq A_n \supseteq \ldots$ is a *decreasing sequence* of sets $A_n \in \mathfrak{A}$, then $\bigcap_{n=1}^{\infty} A_n \in \mathfrak{A}$.

In words, a monotone class is a family $\mathfrak{A}$ of subsets of X that is closed under the operations of taking unions of countable increasing sequences of members of $\mathfrak{A}$ and of taking intersections of countable decreasing sequences of members of $\mathfrak{A}$.

Definition 2.1.4 A *Borel field* is a field $\mathfrak{A}$ that is also a monotone class.

Theorem 2.1.1 *A family* $\mathfrak{A}$ *of subsets of* X *is a Borel field if and only if it is closed under the operations of complementation and of taking unions of countable sequences of elements of* $\mathfrak{A}$.

Remark 2.1.1 Because of Theorem 2.1.1, a Borel field is often called either a σ-*field* or a σ-*algebra*. The letter σ connotes set-theoretic summation, or union, of arbitrary countable sequences.

EXERCISES

2.3 Prove Theorem 2.1.1.

2.4 Let $f : X \to Y$ be any function from one set into another. Let $\mathfrak{A} \subseteq \mathfrak{P}(Y)$ be any σ-algebra of subsets of Y. Prove: The family

$$f^{-1}(\mathfrak{A}) = \left\{ f^{-1}(A) \mid A \in \mathfrak{A} \right\} \subseteq \mathfrak{P}(X)$$

is also a σ-algebra.

2.5 Show that the field of elementary sets in $[0, 1)$, as defined in Example 2.1, is *not* a Borel field. Hint: consider

$$\bigcup_{n=1}^{\infty} \left[\frac{1}{n}, 1 \right).$$

Definition 2.1.5 Let $\mathfrak{A}$ be any family of subsets of X. Define the *Borel field*, $\mathbb{B}(\mathfrak{A})$, *generated* by $\mathfrak{A}$, to be the intersection of all the σ-fields that contain $\mathfrak{A}$, and define the *monotone class*, $\mathbb{M}(\mathfrak{A})$, *generated by* $\mathfrak{A}$ to be the intersection of all the monotone classes that contain $\mathfrak{A}$. An element of $\mathbb{B}(\mathfrak{A})$ is called a *Borel set*.

Whether or not a subset S of an abstract set X is a Borel set depends upon what field $\mathfrak{A}$ is used to determine the Borel field $\mathbb{B}(\mathfrak{A})$. For certain important examples, such as the real line $\mathbb{R}$ or Euclidean space $\mathbb{R}^n$, the choice of $\mathfrak{A}$ will be the field of *elementary sets* defined in Examples 2.2–2.4.

EXERCISES

2.6   Let $\mathfrak{A}$ be any family of subsets of X. Show that $\mathfrak{A}$ is contained in each of the following: a field, a Borel field, and a monotone class.

2.7   Give an example of an infinite set X and a family $\mathfrak{A} \subseteq \mathfrak{P}(X)$ for which $\mathbb{M}(\mathfrak{A})$ is *not* a field of sets.

The following theorem will be useful.

Theorem 2.1.2 *If $\mathfrak{A}$ is a field of subsets of X, then $\mathbb{M}(\mathfrak{A}) = \mathbb{B}(\mathfrak{A})$.*

Proof: Since each Borel field is a monotone class, it follows that

$$\mathbb{M}(\mathfrak{A}) \subseteq \mathbb{B}(\mathfrak{A}).$$

This much would be true even if $\mathfrak{A}$ had not been a field.

Thus it will suffice to prove that $\mathbb{M}(\mathfrak{A}) \supseteq \mathbb{B}(\mathfrak{A})$. Because $\mathbb{M}(\mathfrak{A})$ is a monotone class, it will suffice to prove that $\mathbb{M}(\mathfrak{A})$ is a field (and consequently a Borel field as well). Thus it will suffice to show that if A and B lie in $\mathbb{M}(\mathfrak{A})$, we must have

   i. $A \cap B \in \mathbb{M}(\mathfrak{A})$ and

   ii. $X \backslash A \in \mathbb{M}(\mathfrak{A})$.

We present the rather subtle proof of these two requirements as follows. We begin by fixing temporarily $A \in \mathbb{M}(\mathfrak{A})$ and defining

$$\mathfrak{B}_A = \{B \in \mathfrak{P}(X) \mid A \cap B \in \mathbb{M}(\mathfrak{A})\}. \tag{2.1}$$

We will prove that $\mathfrak{B}_A$ must be a monotone class. Suppose that B_n is an increasing sequence of elements of $\mathfrak{B}_A$. We wish to show that $\bigcup_{n=1}^{\infty} B_n \in \mathfrak{B}_A$. But

$$A \cap \left(\bigcup_{n=1}^{\infty} B_n \right) = \bigcup_{n=1}^{\infty} (A \cap B_n) \in \mathbb{M}(\mathfrak{A})$$

since the latter set is a monotone class and since $A \cap B_n$ is an increasing sequence of sets in $\mathbb{M}(\mathfrak{A})$. The reader should do Exercise 2.8 to prove that $\mathfrak{B}_A$ is a monotone class.

   I. Now that we know that $\mathfrak{B}_A$ is a monotone class, to complete the proof of (i) we need to prove that $\mathfrak{B}_A \supseteq \mathfrak{A}$ so that we will know that $\mathfrak{B}_A \supseteq \mathbb{M}(\mathfrak{A})$.

     (a) We start with a special case by *restricting* the set A to be in $\mathfrak{A}$, fixed temporarily. That $\mathfrak{B}_A \supseteq \mathfrak{A}$ follows in this special case from the fact that $\mathfrak{A}$ is a field. Since we know already that $\mathfrak{B}_A$ is a monotone class, and since in subcase (a) we know that $\mathfrak{B}_A \supseteq \mathfrak{A}$, it follows that $\mathfrak{B}_A \supseteq \mathbb{M}(\mathfrak{A})$. Thus we have shown that $\mathbb{M}(\mathfrak{A})$ is closed under the operation of taking intersections with all $A \in \mathfrak{A}$.

(b) For the full requirement (i), we relax the initial restriction and allow $A \in \mathbb{M}(\mathfrak{A})$ to be fixed arbitrarily as in Equation (2.1), and we let $\mathfrak{B}_A$ be defined still by Equation (2.1). And now *we know from subcase* (a) that $\mathfrak{B}_A \supseteq \mathfrak{A}$. Since $\mathfrak{B}_A$ is a monotone class, it follows that $\mathfrak{B}_A \supseteq \mathbb{M}(\mathfrak{A})$, which is therefore closed under the operation of intersection.

II. Let

$$\mathfrak{A}' = \{A \in \mathfrak{P}(X) \mid X \backslash A \in \mathbb{M}(\mathfrak{A})\}.$$

Since $\mathfrak{A}' \supseteq \mathfrak{A}$ because $\mathfrak{A}$ is a field, it suffices to show that $\mathfrak{A}'$ is a monotone class and therefore contains $\mathbb{M}(\mathfrak{A})$. So we let A_n be an increasing sequence of elements of $\mathfrak{A}'$ and we wish to show that $\bigcup_{n=1}^{\infty} A_n \in \mathfrak{A}'$. But

$$X \backslash \left(\bigcup_{n=1}^{\infty} A_n \right) = \bigcap_{n=1}^{\infty} (X \backslash A_n) \in \mathbb{M}(\mathfrak{A}),$$

because $\bigcap_{n=1}^{\infty} (X \backslash A_n)$ is the intersection of a decreasing sequence of sets in the monotone class $\mathbb{M}(\mathfrak{A})$.

Exercise 2.9 will complete the proof of Theorem 2.1.2.

∎

EXERCISES

2.8 To complete the proof that $\mathfrak{B}_A$ is a monotone class, the reader should prove that $\mathfrak{B}_A$, as defined in Equation (2.1), is closed under the operation of taking intersections of decreasing sequences in $\mathfrak{B}_A$.

2.9 To complete the proof of Theorem 2.1.2, the reader should prove that $\mathfrak{A}'$ is closed under the operation of intersection of decreasing sequences of sets.

2.10 Give an example of a set X and a subset $\mathfrak{A} \subset \mathfrak{P}(X)$ that is *not* a field, yet $\mathbb{M}(\mathfrak{A}) = \mathbb{B}(\mathfrak{A})$.

Remark 2.1.2 The reader should not be surprised if the proof of the preceding theorem seems quite abstract. We will study a variety of theorems to help us understand which sets are Borel sets or measurable sets in familiar examples. But we will also learn theorems that show us that our intuition regarding Borel sets and measurable sets is limited. The reason that the proof of Theorem 2.1.2 does not give us a sense of intuition about the family of Borel sets is that $\mathbb{B}(\mathfrak{A})$ is defined as the intersection of a huge family of σ-fields containing $\mathfrak{A}$. The reader has a very easy example of a σ-field that contains every $\mathfrak{A}$, but the totality of such σ-fields is immense.

Be alert to the different contexts in which the term *Borel set* is used. In this book, in the context of the real line, a Borel set means a set that is in the Borel field generated by the field of finite unions of left-closed, right-open intervals. We will see that this is the same as the Borel field generated by the family of open sets or by the family of closed sets. In the context of an abstract field $\mathfrak{A}$ within the power set

$\mathfrak{P}(X)$ of an abstract set X, the term *Borel set* refers to any element of the Borel field $\mathbb{B}(\mathfrak{A})$ generated by $\mathfrak{A}$.

■ **EXAMPLE 2.2**

Let $\mathfrak{E}$ be the field of elementary subsets of the unit interval $[0, 1)$, as in Example 2.1. The elements $E \in \mathbb{B}(\mathfrak{E})$ are called the *Borel sets of the unit interval*. Sometimes $\mathbb{B}(\mathfrak{E})$ is denoted as $\mathfrak{B}(X)$ where $X = [0, 1)$, so that the notation connotes the Borel subsets of the unit interval with respect to the standard field of elementary sets.

The reader should note that every interval within $[0, 1)$ is a Borel set, regardless of whether it is half-closed, closed, or open.[7] But this scarcely scratches the surface of the family of Borel sets, as the reader will see later in Example 3.11.

Having observed earlier, in Example 2.1, that there is an immediate extension of the concept of an elementary set to any finite interval $[-N, N) \subset \mathbb{R}$, we can extend the concept of a Borel set to $[-N, N)$ as well. Moreover, a subset S of the real line $\mathbb{R}$ itself is called Borel provided that $S \cap [-N, N)$ is Borel for each $N \in \mathbb{N}$.

■ **EXAMPLE 2.3**

Let

$$X = [0, 1) \times [0, 1) = [0, 1)^2,$$

the unit square in the plane $\mathbb{R}^2$. Let I_k and J_k denote left-closed, right-open subintervals of $[0, 1)$ for each $k = 1, \ldots, n$. Let $\mathfrak{E}$ denote the family of all subsets of X of the form

$$E = \bigcup_{k=1}^{n} I_k \times J_k,$$

where $n \in \mathbb{N}$ is arbitrary but finite and varying, depending on the choice of $E \in \mathfrak{E}$.

The reader should be able to generalize this example to cover elementary sets and Borel sets in $X_N^2 = [-N, N) \times [-N, N)$ for each $N \in \mathbb{N}$. A subset S of $\mathbb{R}^2$ is called a *Borel set in* $\mathbb{R}^2$ if $S \cap X_N^2$ is Borel for each $N \in \mathbb{N}$.

EXERCISE

2.11   Prove that the set $\mathfrak{E}$ of Example 2.3 is a field.

The field $\mathfrak{E}$ of Example 2.3 is called the *field of elementary sets in the unit square*, and $\mathbb{B}(\mathfrak{E})$ is called the *field of Borel sets in the unit square*, also denoted as $\mathbb{B}\left([0, 1)^2\right)$.

---

[7]This is true because each singleton set, $\{x\}$, is a Borel set, as the reader should prove easily.

■ **EXAMPLE 2.4**

Let

$$X = [0, 1)^k,$$

a k-fold Cartesian product of copies of the unit interval. The set X is called the *unit cube* in $\mathbb{R}^k$. Let $\mathfrak{E}$ denote the field of all finite unions of k-dimensional boxes of the form

$$E = I_1 \times \cdots \times I_k,$$

where each I_j is an interval of the form $[a_j, b_j) \subseteq [0, 1)$. Then the set $\mathbb{B}(\mathfrak{E})$ is called the *field of Borel sets of the unit cube* in $\mathbb{R}^k$.

We can extend the concepts of elementary set and Borel set to any cube $[-N, N)^k$ for $N \in \mathbb{N}$. And a subset $S \subseteq \mathbb{R}^k$ is called a Borel set provided that $S \cap [-N, N)^k$ is a Borel set for each $N \in \mathbb{N}$.

■ **EXAMPLE 2.5**

Here we define Borel sets in the product of infinitely many copies of the unit interval $[0, 1)$, the *Tychonoff cube* X of arbitrary infinite dimension.[8] Let Γ be any arbitrary (infinite) index set. (We do not assume that Γ is countable.) For each $\gamma \in \Gamma$, let $S_\gamma = [0, 1)$, the unit interval. According to the Axiom of Choice [16], there exists a function x on Γ such that for each $\gamma \in \Gamma$ we have $x(\gamma) \in S_\gamma$. Thus the function x *chooses* one element from each set S_γ. We define

$$X = \prod_{\gamma \in \Gamma} S_\gamma = \{x \mid x(\gamma) \in S_\gamma \ \forall \gamma \in \Gamma\}.$$

The Axiom of Choice tells us that the Tychonoff cube

$$X = \prod_{\gamma \in \Gamma} S_\gamma \neq \varnothing.$$

Intuitively, we think of $x(\gamma)$ as representing the γth coordinate of the point $x \in X$.

Now let Δ be any *finite* subset of Γ. For each $\gamma \in \Delta$ we specify some subinterval $[a_\gamma, b_\gamma) \subseteq S_\gamma$. By a *cylinder set* in X we mean a set $R \subseteq X$ such that $x \in R$ if and only if

$$x(\gamma) \in \begin{cases} [a_\gamma, b_\gamma) & \text{if } \gamma \in \Delta, \\ S_\gamma & \text{if } \gamma \in \Gamma \backslash \Delta. \end{cases}$$

This means that a cylinder set R is defined by restricting only some finite collection of coordinates to particular subintervals of S_γ. We define an *elementary*

---

[8]It is common to take the Tychonoff cube to be an arbitrary product of copies of the closed unit interval $[0, 1]$, so that the Tychonoff cube will be compact. The set X defined here is a subset of the compact cube.

set in X to be the union

$$\bigcup_{k=1}^{n} R_k$$

of finitely many cylinder sets in X. The set $\mathfrak{E}$ of elementary sets is a field, and $\mathbb{B}(\mathfrak{E})$ is the set of Borel sets in the Tychonoff cube X.

Remark 2.1.3 The discussion up to this point has presented what may be called an *external* construction of the Borel sets corresponding to a field of elementary sets. The word *external* connotes the fact that $\mathbb{B}(\mathfrak{E})$ is defined as the intersection of all Borel fields that *contain* $\mathfrak{E}$. For those who have studied transfinite ordinal numbers and transfinite induction [16], we sketch briefly here an *internal* construction of $\mathbb{B}(\mathfrak{E})$. The interested reader can consult reference [8] for details.

For the case of Euclidean space $\mathbb{R}^n$ or the cube therein, the procedure may be described briefly as follows. We begin with a field $\mathfrak{A}$ that happens not to be a Borel field, meaning that it is not a monotone class. So we construct a larger field by tacking on the unions and intersections of all increasing and decreasing nested sequences of sets from $\mathfrak{A}$. The resulting family is called $\mathfrak{A}_1$. The same procedure is applied to subsets of $\mathfrak{A}_1$ in order to augment $\mathfrak{A}_1$ so as to produce $\mathfrak{A}_2$. We do this for each finite ordinal number n, and then for the first infinite ordinal, ω, we define the union of all the infinitely many families already constructed to be $\mathfrak{A}_\omega$. Then we apply the original procedure to $\mathfrak{A}_\omega$ to produce $\mathfrak{A}_{\omega+1}$, etc. At each nonlimit ordinal number we apply the original procedure, and at each limit ordinal, we take the union over all its predecessors and repeat the process. When this has been done for all ordinal numbers $\lambda < \Omega$, the first *uncountable* ordinal, the process stops, having reached a Borel field. The constructed field is the same one, called $\mathbb{B}(\mathfrak{A})$, that we obtained much more easily by the external method.

A good example of an interesting Borel set that the reader will meet is the Cantor set, to be defined in Example 3.11. It is a common but serious error to imagine that a subset of the interval $[0, 1)$ is a Borel set if and only if it is a union or an intersection of countably many intervals, or intervals of the specified type. This is false, and the elaborate transfinite induction we have described suggests this fact.

2.2   ADDITIVE MEASURES

Let $\mathbb{R}^* = \mathbb{R} \cup \{\infty, -\infty\}$ denote the *extended real number system*, and

$$[0, \infty] = \{x \in \mathbb{R}^* \mid 0 \leqslant x \leqslant \infty\}.$$

Let X be an abstract set and $\mathfrak{A}$ a field of subsets of X.

Definition 2.2.1 A function $\mu : \mathfrak{A} \to [0, \infty]$ is called a *finitely additive measure* if and only if

  i. If $A = \bigcup_{i=1}^{n} A_i$ is the union of finitely many mutually disjoint sets $A_i \in \mathfrak{A}$, then

$$\mu(A) = \sum_{i=1}^{n} \mu(A_i).$$

ii. $\mu(\varnothing) = 0$.

If $\mu(X) < \infty$, then μ is called *finite*: Otherwise, μ is called *infinite*. The measure μ is called *approximately finite* if, for all $A \in \mathfrak{A}$ with $\mu(A) = \infty$ and for all real numbers $M > 0$, there exists $B \subset A$ such that $M < \mu(B) < \infty$.

■ **EXAMPLE 2.6**

Let X be any infinite set and $\mathfrak{A} = \mathfrak{P}(X)$, the power set of X. Let $\mu(A) = 0$ if $A = \varnothing$ and $\mu(A) = \infty$ if $A \neq \varnothing$. Then μ is finitely additive but it is not approximately finite. Next, define a finitely additive measure ν by letting $\nu(A)$ be the number of elements in A if A is finite, and letting $\nu(A) = \infty$ if A is infinite. Then ν is approximately finite.

■ **EXAMPLE 2.7**

Let $X = [0, 1)$, the unit interval, and let $\mathfrak{E}$ be the field of elementary sets in the unit interval. Thus if $E \in \mathfrak{E}$, we can express

$$E = \bigcup_{i=1}^{n} [a_i, b_i),$$

a *disjoint* union of finitely many left-closed right-open intervals.[9] Suppose we are given a *nondecreasing* function $f : [0, 1] \to \mathbb{R}$ for which $f(0) = 0$. We wish to define

$$\mu(E) = \sum_{i=1}^{n} [f(b_i) - f(a_i)]$$

and to show that μ is a finite, finitely additive measure on $\mathfrak{E}$. The work is left to Exercise 2.13.

Because finitely additive measures are not sufficient for the purposes of analysis, we make the following definition.

Definition 2.2.2 A *countably additive measure on a field* $\mathfrak{A}$ is a finitely additive measure μ with the following property. If $A_i \in \mathfrak{A}$ is an infinite sequence of *disjoint* sets for which

$$A = \bigcup_{i=1}^{\infty} A_i \in \mathfrak{A},$$

then we have

$$\mu(A) = \sum_{i=1}^{\infty} \mu(A_i).$$

---

[9]Here we understand the convention to be that $[a, b) = \{x \in \mathbb{R} \mid a \leqslant x < b\}$, with $a \leqslant b$. Note that $[a, a) = \varnothing$.

Note that we do not require $\mathfrak{A}$ to be a σ-field in the preceding definition.

Definition 2.2.3 A measure μ that is defined on a field $\mathfrak{A} \subseteq \mathfrak{P}(X)$ is called σ-*finite* if $X = \bigcup_{i \in \mathbb{N}} X_i$, with $\mu(X_i) < \infty$ for each $i \in \mathbb{N}$.

Note that a finite measure is σ-finite as well, because $X = \bigcup_{i \in \mathbb{N}} X$, the union of an infinite sequence of copies of X itself.

EXERCISES

2.12   Show that condition (ii) of Definition 2.2.1 can be replaced by the following condition: (ii') $\mu(\varnothing) < \infty$.

2.13   The decomposition of E used to define μ in Example 2.7 is not unique. Prove that μ is nonetheless well defined. Prove also that there is a bijection between the set of all finite, finitely additive measures on $\mathfrak{E}$ in that example and the set of all nondecreasing real-valued functions ϕ on $[0, 1]$ for which $\phi(0) = 0$.

2.14   Prove that every infinite σ-finite countably additive measure μ is approximately finite.

2.15   Let X be any set, and let $\mathfrak{A} = \mathfrak{P}(X)$, the power set of X. Define the *counting measure* $\nu(E)$ to be the cardinality of E for each *finite* set $E \subseteq X$. Let $\nu(E) = \infty$ for each infinite subset of X.

   a) Prove that $\mathfrak{A}$ is a σ-field in $\mathfrak{P}(X)$, and that ν is countably additive on $\mathfrak{A}$. [10]

   b) Prove that ν is a σ-finite measure on $\mathfrak{A}$ if and only if X is at most a countably infinite set.

2.16   Give an example of a finitely additive measure μ on the set $\mathbb{N}$ of all natural numbers such that μ is *not* countably additive. [11]

2.3   CARATHÉODORY OUTER MEASURE

In this section we define still another type of set function, called a *Carathéodory outer measure*. We will show that for this type of set function, there exists a special class of sets, called *measurable sets*, on which the set function is a countably additive measure. To begin with, we do not assume that the Carathéodory outer measure is even finitely additive, but rather only that it is what is called *countably subadditive*. (In the next section, we will determine the necessary and sufficient conditions for defining a Carathéodory outer measure that extends a given finitely additive measure on a given field of sets.)

---

[10]This exercise establishes that the triplet $(X, \mathfrak{A}, \nu)$ is a *measure space*, in the sense that will be presented in Definition 3.3.1.

[11]An example of a finitely additive measure, defined on suitable subsets of the *Euclidean plane*, that fails to be countably additive will be provided by Exercise 3.28.

Definition 2.3.1 Let X be a non-$\emptyset$ set, and let

$$\mu : \mathfrak{P}(X) \to [0, \infty].$$

Then μ is called a *Carathéodory outer measure* if and only if it satisfies the following three conditions:

i. $\mu(\emptyset) = 0$.

ii. $A \subset B \implies \mu(A) \leqslant \mu(B)$, which is called *monotonicity*.[12]

iii. $A \subseteq \bigcup_{i=1}^{\infty} A_i \implies \mu(A) \leqslant \sum_{i=1}^{\infty} \mu(A_i)$, which is called *countable subadditivity*.

Next, we define the σ-field $\mathfrak{A}$ of subsets $A \in \mathfrak{P}(X)$ on which a Carathéodory outer measure will turn out to be countably additive.

Definition 2.3.2 Let μ be a Carathéodory outer measure defined on $\mathfrak{P}(X)$. A subset $A \subseteq X$ is called μ-*measurable*[13] if, and only if, *for all* $W \in \mathfrak{P}(X)$ we have

$$\mu(W) = \mu(W \cap A) + \mu(W \cap A^c),$$

where $A^c = X \backslash A$, the complement of A.

We observe that without the condition in this definition, all we would have known is that

$$\mu(W) \leqslant \mu(W \cap A) + \mu(W \cap A^c).$$

We could describe the definition in words as stating that A is μ-measurable if, and only if, *A and its complement split every set W additively with respect to the Carathéodory outer measure* μ. Next, we will show that the family of all μ-measurable sets A is a σ-field on which μ is a countably additive measure.

Theorem 2.3.1 *Let μ be a Carathéodory outer measure defined on $\mathfrak{P}(X)$. Then the family $\mathfrak{A}$ of all μ-measurable subsets of X is a σ-field, and μ is countably additive on $\mathfrak{A}$.*

Proof: It will suffice to prove the following three statements:

i. $A \in \mathfrak{A}$ implies that $A^c = X \backslash A \in \mathfrak{A}$.

ii. A and B in $\mathfrak{A}$ implies that $A \cap B \in \mathfrak{A}$.

iii. If the *mutually disjoint* sets A_i lie in $\mathfrak{A}$ for all $i \in \mathbb{N}$, then

$$A = \bigcup_{i=1}^{\infty} A_i \in \mathfrak{A}$$

---

[12]This property is redundant, but is stated here for emphasis because it is important. See Exercise 2.17.
[13]For a simpler but equivalent definition in the special case in which $\mu(X) < \infty$, see Theorem 2.4.2.

and

$$\mu(A) = \sum_{i \in \mathbb{N}} \mu(A_i).$$

Note that properties (i) and (ii) will suffice to establish that $\mathfrak{A}$ is a field. The field property is important for the proof that property (iii) suffices to establish closure of $\mathfrak{A}$ under countable unions even for nondisjoint sequences of sets in $\mathfrak{A}$.[14] Otherwise, one might think that property (ii) is superfluous in light of Theorem 2.1.1.

We begin by proving (i). But this is immediate since if A is μ-measurable, we know for each $W \in \mathfrak{P}(X)$ that

$$\mu(W) = \mu(W \cap A) + \mu(W \cap A^c),$$

and this implies that A^c is measurable as well by the symmetry of the definition.

Next, we prove part (ii). We must show that if A and B are μ-measurable, then

$$\mu(W) = \mu(W \cap (A \cap B)) + \mu(W \cap (A \cap B)^c),$$

where $(A \cap B)^c = A^c \cup B^c$. We know that

$$\begin{aligned}
\mu(W) &= \mu(W \cap A) + \mu(W \cap A^c) \\
&= \mu(W \cap (A \cap B)) + \mu(W \cap (A \cap B^c)) + \mu(W \cap A^c) \\
&\geq \mu(W \cap (A \cap B)) + \mu(W \cap (A \cap B)^c),
\end{aligned}$$

because we have

$$(A \cap B)^c = A^c \cup B^c = A^c \cup (A \cap B^c).$$

This means that the *subadditive* measure μ is also *superadditive* when we split W using the pair $A \cap B$ and $(A \cap B)^c$. Hence W is split additively and $A \cap B$ is measurable.

Finally, we need to prove part (iii). Keep in mind that the sets A_i are *mutually disjoint*. For arbitrary W we have

$$\begin{aligned}
\mu(W) &= \mu(W \cap A_1) + \mu(W \cap A_1^c) \\
&= \mu(W \cap A_1) + \mu(W \cap A_1^c \cap A_2) + \mu(W \cap A_1^c \cap A_2^c) \\
&= \mu(W \cap A_1) + \mu(W \cap A_2) + \mu(W \cap (A_1 \cup A_2)^c) \\
&= \cdots = \sum_{i=1}^{n} \mu(W \cap A_i) + \mu\left(W \cap \left(\bigcup_{i=1}^{n} A_i\right)^c\right) \\
&\geq \sum_{i=1}^{n} \mu(W \cap A_i) + \mu\left(W \cap \left(\bigcup_{i=1}^{\infty} A_i\right)^c\right)
\end{aligned}$$

[14]See Exercise 2.18.

by monotonicity of μ. Since this is true for all n, we have also that

$$\mu(W) \geqslant \sum_{i=1}^{\infty} \mu(W \cap A_i) + \mu\left(W \cap \left(\bigcup_{i=1}^{\infty} A_i\right)^c\right) \qquad (2.2)$$

Also, by subadditivity,

$$\mu(W \cap A) = \mu\left(\bigcup_{i=1}^{\infty}(W \cap A_i)\right) \leqslant \sum_{i=1}^{\infty} \mu(W \cap A_i).$$

Hence Equation (2.2) tells us that

$$\mu(W) \geqslant \mu(W \cap A) + \mu(W \cap A^c),$$

and then subadditivity of μ implies that

$$\mu(W) = \mu(W \cap A) + \mu(W \cap A^c).$$

It follows that $A \in \mathfrak{A}$, so that $\mathfrak{A}$ is a σ-field by virtue of Exercise 2.18. And by subadditivity applied to Inequality (2.2), we have

$$\mu(W) = \sum_{i=1}^{\infty} \mu(W \cap A_i) + \mu\left(W \cap \left(\bigcup_{i=1}^{\infty} A_i\right)^c\right). \qquad (2.3)$$

Applying Equation (2.3) to A itself in place of W, we see that

$$\mu\left(\bigcup_{i=1}^{\infty} A_i\right) = \sum_{i=1}^{\infty} \mu(A_i),$$

because $\mu(\varnothing) = 0$.

The proof will be completed by Exercise 2.18. ∎

The preceding theorem gives us a way of obtaining a countably additive measure on an identifiable σ-field of measurable sets, provided we can come up with a suitable Carathéodory outer measure. The Hopf Extension Theorem of the next section will make use of this theorem to enable us to start with a finitely additive measure on a field $\mathfrak{A}$ (of elementary sets) and extend that measure to a countably additive measure defined at least on the set $\mathbb{B}(\mathfrak{A})$, the Borel field generated by $\mathfrak{A}$.

EXERCISES

2.17 In Definition 2.3.1, show that properties (i) and (iii) imply (ii). Moreover, μ is also finitely subadditive.

2.18 Show that, in part (iii) of the proof of Theorem 2.3.1, closure under countable disjoint unions implies closure under countable arbitrary unions.

2.4   E. HOPF'S EXTENSION THEOREM

Suppose we are given an abstract set X, a field $\mathfrak{A} \subseteq \mathfrak{P}(X)$, and a *finitely* additive measure μ defined on $\mathfrak{A}$. We ask under what conditions it will be possible to extend μ to a *countably* additive measure on the σ-field $\mathbb{B}(\mathfrak{A})$ and whether such an extension will be unique. We will call $\mathfrak{A}$ a field of *abstract elementary sets*. The field $\mathbb{B}(\mathfrak{A})$ will be called a field of *abstract Borel sets*.

In a typical application of the Hopf Extension Theorem, we will have a specific set X for the underlying space and a specific field $\mathfrak{A}$ that we will use for *elementary sets* for that space. Usually there will be some naturally defined finitely additive measure μ on $\mathfrak{A}$. For example, in the next chapter, we will pick the field of elementary sets to be the one that is generated by the left-closed, right-open intervals in the real line, and we will use Euclidean length to determine the finitely additive measure μ. Then the Hopf Extension Theorem will be used to produce Lebesgue measure. Different choices of elementary sets and of finitely additive measures of elementary sets may be made to produce different countably additive measures on the real line or elsewhere.

Recall that, by Definition 2.2.2, a finitely additive measure μ is said to be countably additive *on* a field $\mathfrak{A}$ if and only if for each sequence of *disjoint* sets $A_i \in \mathfrak{A}$, such that $A = \bigcup_{i \in \mathbb{N}} A_i \in \mathfrak{A}$, we have

$$\mu(A) = \sum_{i=1}^{\infty} \mu(A_i).$$

Theorem 2.4.1 *Let μ be a finitely additive measure on a field $\mathfrak{A} \subseteq \mathfrak{P}(X)$. In order that there exist a countably additive extension μ^* of μ defined on a σ-field $\mathfrak{A}^* \supseteq \mathfrak{A}$, it is necessary and sufficient that μ be countably additive on $\mathfrak{A}$.* [15] *Moreover, if $(X, \mathfrak{A}, \mu)$ is σ-finite, then the extension μ^* is unique.*

Proof: We prove first the *necessity* of the condition. We assume that there exists a countably additive extension μ^* of μ. Since it is an *extension*, we must have $\mu^*(A) = \mu(A)$ for all $A \in \mathfrak{A}$. Thus, if

$$A = \bigcup_{i \in \mathbb{N}} A_i \in \mathfrak{A},$$

with the sets $A_i \in \mathfrak{A}$ for all i being mutually disjoint, then

$$\mu(A) = \mu^*(A) = \sum_{i \in \mathbb{N}} \mu^*(A_i) = \sum_{i \in \mathbb{N}} \mu(A_i).$$

Thus μ is countably additive on $\mathfrak{A}$.

Next, we will prove the *sufficiency* of the condition. We will define the *function* μ^* on the power set $\mathfrak{P}(X)$ by

$$\mu^*(B) = \inf \left\{ \sum_{i=1}^{\infty} \mu(B_i) \,\middle|\, B \subseteq \bigcup_{i \in \mathbb{N}} B_i, \; B_i \in \mathfrak{A} \; \forall i \in \mathbb{N} \right\}, \qquad (2.4)$$

---

[15]The reader should note that such a σ-field $\mathfrak{A}^*$ must contain $\mathbb{B}(\mathfrak{A})$ as well.

where the infimum is taken over all countable coverings of B by sets B_i in $\mathfrak{A}$. Recall that every field $\mathfrak{A}$ includes as an element the whole set X, so that the family of countable coverings is nonempty. Note that in Equation (2.4), we do *not* require the sets B_i to be disjoint. We will prove first that μ^* is a Carathéodory outer measure. There are three properties to be confirmed:

1. $\mu^*(\varnothing) = 0$. This is easy to confirm, since $\varnothing \in \mathfrak{A}$, so that it can be covered by itself.

2. If $A \subseteq B$, then $\mu^*(A) \leqslant \mu^*(B)$. This follows immediately from the fact that every covering of B covers A as well, and the infimum over a superset will be smaller.

3. If $B \subseteq \bigcup_{i=1}^{\infty} B_i$, we must show that $\mu^*(B) \leqslant \sum_{i=1}^{\infty} \mu^*(B_i)$. (Here, B and B_i are arbitrary subsets of X: They need need not be elements of $\mathfrak{A}$.)

   This will be immediate if
   $$\sum_{i=1}^{\infty} \mu^*(B_i) = \infty.$$

So we will assume that $\sum_{i=1}^{\infty} \mu^*(B_i) < \infty$ and we will prove (3) as follows. Let $\epsilon > 0$. We can cover each set $B_n \subseteq \bigcup_{i=1}^{\infty} B_{n,i}$ in such a way that each $B_{n,i} \in \mathfrak{A}$ and
$$\sum_{i=1}^{\infty} \mu(B_{n,i}) < \mu^*(B_n) + \frac{\epsilon}{2^n}.$$
Since the family $\{B_{n,i} \mid i \in \mathbb{N}, n \in \mathbb{N}\}$ is countable[16] and covers B, it follows from absolute convergence that
$$\mu^*(B) \leqslant \sum_{n \in \mathbb{N}, i \in \mathbb{N}} \mu(B_{n,i})$$
$$\stackrel{(i)}{=} \sum_{n=1}^{\infty} \left(\sum_{i=1}^{\infty} \mu(B_{n,i}) \right)$$
$$\leqslant \sum_{n=1}^{\infty} \left(\mu^*(B_n) + \frac{\epsilon}{2^n} \right)$$
$$= \left(\sum_{n=1}^{\infty} \mu^*(B_n) \right) + \epsilon$$

---

[16]If C is any countable collection of numbers, then the concept of summation of that countable set of nonnegative numbers is as follows:
$$\sum_{x \in C} x = \sup \left\{ \sum_{x \in F} x \,\middle|\, F \subseteq C, \; \#F < \infty \right\},$$
meaning that F is required to be finite. In advanced calculus, this is treated also for sums of real numbers of arbitrary sign, provided that the sum as defined above is finite for the absolute values. See, for example, [20].

for all $\epsilon > 0$. Here we use the fact that the union of countably many countable sets is countable. For the equality (i) we have used the properties of absolutely convergent *double-series*[17] with respect to summation first by rows and then by columns, or vice versa. This is sufficient to prove (3).

Next, we must show that μ^* is an extension of μ and that $\mathfrak{A}$ is contained within the family of μ^*-measurable sets.

a. We will show that for all $A \in \mathfrak{A}$, we have $\mu^*(A) = \mu(A)$. What is clear is that since $A \subseteq A \cup \varnothing \cup \varnothing \cup \ldots$ we must have $\mu^*(A) \leqslant \mu(A)$. In order to prove the opposite inequality, we will make use of the hypothesis of countable additivity *on* $\mathfrak{A}$, as in Definition 2.2.2. So, suppose we have a countable covering of A: $A \subseteq \bigcup_{i=1}^{\infty} A_i$, with each $A_i \in \mathfrak{A}$. We can construct from this covering a countable *disjoint* covering by sets $B_i \in \mathfrak{A}$ defined as follows: Let $B_1 = A_1 \cap A$, $B_2 = (A_2 \cap A)\backslash B_1$, and in general

$$B_n = (A_n \cap A)\backslash \left(\bigcup_{i=1}^{n-1} B_i \right).$$

Now A is covered by a countable disjoint family of sets $B_i \in \mathfrak{A}$. Countable additivity on $\mathfrak{A}$ tells us that

$$\mu(A) = \sum_{i=1}^{\infty} \mu(B_i) \leqslant \sum_{i=1}^{\infty} \mu(A_i)$$

for all countable covers by sets A_i, where we are using the monotonicity of μ. Therefore, $\mu(A) \leqslant \mu^*(A)$. Hence $\mu(A) = \mu^*(A)$.

b. Since we know that μ^* is a Carathéodory outer measure, we can denote by $\mathfrak{A}^*$ the family of all μ^*-measurable sets, as in Definition 2.3.2. We know that $\mathfrak{A}^*$ is a σ-field. If we can show that $\mathfrak{A} \subseteq \mathfrak{A}^*$, then we will know that μ^* is an extension of μ, and it will follow also that $\mathbb{B}(\mathfrak{A}) \subseteq \mathfrak{A}^*$. So, we let $A \in \mathfrak{A}$, and we let $W \in \mathfrak{P}(X)$. According to Definition 2.3.2, we must prove that

$$\mu^*(W) = \mu^*(W \cap A) + \mu^*(W \cap A^c).$$

Because of subadditivity of μ^*, it will suffice to prove that

$$\mu^*(W) \geqslant \mu^*(W \cap A) + \mu^*(W \cap A^c).$$

---

[17] A *double series* is a sum of the form $\sum_{i \in \mathbb{N}, j \in \mathbb{N}} x_{i,j}$, and absolute convergence implies that

$$\sum_{i \in \mathbb{N}, j \in \mathbb{N}} x_{i,j} = \sum_{i \in \mathbb{N}} \left(\sum_{j \in \mathbb{N}} x_{i,j} \right),$$

which the reader can find in many advanced calculus texts, such as [20]. This is also a special case of Fubini's Theorem, which is Theorem 6.2.2 in the present text.

This inequality would be immediate if $\mu^*(W) = \infty$, so we will assume that $\mu^*(W) < \infty$.

To this end, we let $\epsilon > 0$. There exist $A_i \in \mathfrak{A}$ such that $W \subseteq \bigcup_{i \in \mathbb{N}} A_i$ and

$$\sum_{i=1}^{\infty} \mu(A_i) < \mu^*(W) + \epsilon.$$

Note that

$$\mu^*(W \cap A) \leqslant \sum_{i=1}^{\infty} \mu(A_i \cap A),$$

and $\mu^*(W \cap A^c) \leqslant \sum_{i=1}^{\infty} \mu(A_i \cap A^c)$. Thus

$$\mu^*(W \cap A) + \mu^*(W \cap A^c) \leqslant \sum_{i=1}^{\infty} \mu(A_i) < \mu^*(W) + \epsilon,$$

using absolute convergence of the series and finite additivity of μ on $\mathfrak{A}$. Since $\epsilon > 0$ is arbitrary, the proof is complete except for uniqueness, which is treated in Exercise 2.21.

∎

Definition 2.4.1 A Carathéodory outer measure μ defined on $\mathfrak{P}(X)$ is called *regular* with respect to a field $\mathfrak{A} \subseteq \mathfrak{P}(X)$ if and only if for each $A \in \mathfrak{P}(X)$ there exists a Borel set $B \in \mathbb{B}(\mathfrak{A})$ such that $A \subseteq B$ and $\mu(A) = \mu(B)$.[18]

We turn our attention next to the questions of the uniqueness and regularity of the extension.

EXERCISES

2.19 Let μ be a *finite, finitely additive* measure [19] on a field $\mathfrak{A} \subseteq \mathfrak{P}(X)$. Prove[20] that μ is countably additive *on* $\mathfrak{A}$ if and only if for each decreasing sequence

$$A_1 \supseteq A_2 \supseteq \ldots \supseteq A_n \supseteq \ldots$$

with $\bigcap_{n=1}^{\infty} A_n = \varnothing$, we have $\lim_{n \to \infty} \mu(A_n) = 0$.

2.20 The Carathéodory outer measure on $\mathfrak{P}(X)$, constructed in Theorem 2.4.1, is regular. Hint: Let $A \in \mathfrak{P}(X)$. For each $n \in \mathbb{N}$ pick

$$B_n = \bigcup_{i=1}^{\infty} A_{n,i} \in \mathbb{B}(\mathfrak{A}),$$

---

[18] See Exercise 2.20.
[19] Finiteness of the measure means that $\mu(X) < \infty$.
[20] Variations on this exercise appear in Exercises 5.20, 5.24, and 8.10.

with each $A_{n,i} \in \mathfrak{A}$, for which $\mu^*(B_n) < \mu^*(A) + \frac{1}{n}$.

2.21   Suppose that *both* μ_1 and μ_2 are countably additive *extensions* of the measure μ from the field $\mathfrak{A}$ to $\mathbb{B}(\mathfrak{A})$. Suppose that μ is countably additive *on* $\mathfrak{A}$.

    a) Suppose that $\mu(X) < \infty$. Prove that $\mu_1(B) = \mu_2(B)$ for all $B \in \mathbb{B}(\mathfrak{A})$. (Hint: Show that the set

$$\mathfrak{B}(\mathfrak{A}) = \{B \in \mathbb{B}(\mathfrak{A}) \mid \mu_1(B) = \mu_2(B)\}$$

is closed under complementation and under taking unions of increasing sequences, making $\mathfrak{B}(\mathfrak{A})$ a monotone class that contains $\mathfrak{A}$. The finiteness of $\mu(X)$ will be helpful for complementation.)

    b) Now replace the hypothesis that $\mu(X) < \infty$ in part (a) with the hypothesis that μ is σ-finite on X. Prove that $\mu_1(B) = \mu_2(B)$ for all $B \in \mathbb{B}(\mathfrak{A})$. (Hint: Use Corollary 2.4.1.)

Theorem 2.4.2 *Let μ be a* regular *Carathéodory outer measure on $\mathfrak{P}(X)$ such that $\mu(X) < \infty$. Then a subset $A \subseteq X$ is μ-measurable if and only if*

$$\mu(X) = \mu(A) + \mu(A^c).$$

Proof: Necessity is immediate, so we will prove sufficiency. We need to prove for each $W \in \mathfrak{P}(X)$ that

$$\mu(W) = \mu(W \cap A) + \mu(W \cap A^c).$$

Because of regularity, there exists a measurable set $V \supseteq W$ such that $\mu(V) = \mu(W)$. Observe that, since V is measurable, we have

$$\begin{aligned}
\mu(A) &= \mu(A \cap V) + \mu(A \cap V^c), \\
\mu(A^c) &= \mu(A^c \cap V) + \mu(A^c \cap V^c).
\end{aligned} \tag{2.5}$$

By hypothesis and by Equations (2.5),

$$\begin{aligned}
\mu(X) = \mu(A) + \mu(A^c) &= \mu(V) + \mu(V^c) \\
&= [\mu(A \cap V) + \mu(A^c \cap V)] + [\mu(A \cap V^c) + \mu(A^c \cap V^c)].
\end{aligned}$$

On the other hand, it follows from subadditivity that

$$\begin{aligned}
\mu(A \cap V) + \mu(A^c \cap V) &\geqslant \mu(V), \\
\mu(A \cap V^c) + \mu(A^c \cap V^c) &\geqslant \mu(V^c).
\end{aligned}$$

It follows that

$$\begin{aligned}
\mu(A \cap V) + \mu(A^c \cap V) &= \mu(V), \\
\mu(A \cap V^c) + \mu(A^c \cap V^c) &= \mu(V^c).
\end{aligned}$$

By the choice of V and by the preceding equations,

$$\mu(W) = \mu(V) = \mu(V \cap A) + \mu(V \cap A^c)$$
$$\geq \mu(W \cap A) + \mu(W \cap A^c)$$
$$\geq \mu(W).$$

Hence

$$\mu(W \cap A) + \mu(W \cap A^c) = \mu(W),$$

and A is measurable.   ∎

2.4.1   Fields, σ-Fields, and Measures Inherited by a Subset

In Definition 3.3.1, we will see that a triplet, $(X, \mathfrak{A}, \mu)$, is called a *measure space*, provided that X is a set, $\mathfrak{A} \subseteq \mathfrak{P}(X)$ is a σ-field, and μ is a countably additive measure defined on $\mathfrak{A}$.

Definition 2.4.2 A triplet $(X, \mathfrak{A}, \mu)$ is a *premeasure space* provided that X is a set, $\mathfrak{A} \subseteq \mathfrak{P}(X)$ is a field, and μ is a finitely additive measure defined on $\mathfrak{A}$.

Thus the Hopf Extension Theorem provides necessary and sufficient conditions for a premeasure space to be extended to a full-fledged measure space. Note that there exist premeasure spaces that cannot be extended to measure spaces. [21]
It is often useful to consider the restriction of a measure μ, given to us in either a premeasure space or a measure space, to a subfield of the power set $\mathfrak{P}(S)$ for some set $S \in \mathfrak{A}$. An especially important instance is the situation in which $(X, \mathfrak{A}, \mu)$ is σ-finite, so that $X = \bigcup_{i \in \mathbb{N}} X_i$, with $\mu(X_i) < \infty$, for each $i \in \mathbb{N}$.

Definition 2.4.3 If $(X, \mathfrak{A}, \mu)$ is any premeasure space, define the premeasure space *inherited* by $S \in \mathfrak{A}$ to be the triplet

$$(S, \mathfrak{A}_S, \mu),$$

where

$$\mathfrak{A}_S = \{A \cap S \mid A \in \mathfrak{A}\} \subseteq \mathfrak{P}(S),$$

and we retain the symbol μ for the restriction to $\mathfrak{A}_S$ of the given measure on $\mathfrak{A}$.

Since $\mathfrak{A}$ is a field, it is clear that $\mathfrak{A}_S \subseteq \mathfrak{A}$, so that μ is defined on $\mathfrak{A}_S$. Moreover, it is easily checked that $\mathfrak{A}_S$ is itself a subfield *of* the field $\mathfrak{P}(S)$, with the understanding that complementation in $\mathfrak{A}_S$ will be with respect to the S, not with respect to X. That is, for $A \in \mathfrak{A}_S$, we define $A^c = S \backslash A$. Again, because $\mathfrak{A}$ is a field, the set S^c also inherits a premeasure space from $(X, \mathfrak{A}, \mu)$.

---

[21] For example, see Exercise 2.16.

Theorem 2.4.3 *If $(X, \mathfrak{A}, \mu)$ is any premeasure space and if $S \in \mathfrak{A}$, then an arbitrary set B belongs to $\mathbb{B}(\mathfrak{A})$ if and only if $B = B_1 \cup B_2$, where $B_1 \in \mathbb{B}(\mathfrak{A}_S)$ and $B_2 \in \mathbb{B}(\mathfrak{A}_{S^c})$.*

We are to understand in this theorem that $\mathbb{B}(\mathfrak{A}_S) \subseteq \mathfrak{P}(S)$. That is, we treat S as the universal set in the definition of the Borel field generated by $\mathfrak{A}_S$.

Proof: We observe that if $\mathfrak{B}$ is a σ-field in $\mathfrak{P}(X)$ containing $\mathfrak{A}$, then both of the following two conditions are met: $\mathfrak{B}_S$ is a σ-field in $\mathfrak{P}(S)$ containing $\mathfrak{A}_S$, and $\mathfrak{B}_{S^c}$ is a σ-field in $\mathfrak{P}(S^c)$ containing $\mathfrak{A}_{S^c}$.

Conversely, if $\mathfrak{B}_1$ is a σ-field in $\mathfrak{P}(S)$ containing $\mathfrak{A}_S$ and if $\mathfrak{B}_2$ is a σ-field in $\mathfrak{P}(S^c)$ containing $\mathfrak{A}_{S^c}$, then we define

$$\mathfrak{B} = \{B_1 \cup B_2 \mid B_1 \in \mathfrak{B}_1, \ B_2 \in \mathfrak{B}_2\}.$$

Then it is clear that $\mathfrak{B}_S = \mathfrak{B}_1$ and $\mathfrak{B}_{S^c} = \mathfrak{B}_2$, and that $\mathfrak{B}$ is a σ-field in $\mathfrak{P}(X)$ containing $\mathfrak{A}$.

The conclusion follows from Definition 2.1.5, in which the Borel field generated by a given field is the intersection of all σ-fields containing the given field. ∎

We note that the preceding theorem does not involve μ and relates only to the pair $(X, \mathfrak{A})$, which is called a *premeasurable space*.

Corollary 2.4.1 *Suppose*

$$X = \bigcup_{i \in \mathbb{N}} X_i,$$

with each $X_i \in \mathfrak{A}$, a field contained in $\mathfrak{P}(X)$. Then $B \in \mathbb{B}(\mathfrak{A})$ if and only if

$$B = \bigcup_{i \in \mathbb{N}} B_i, \ B_i \in \mathbb{B}(\mathfrak{A}_{X_i}) \subseteq \mathfrak{P}(X_i) \ \forall i \in \mathbb{N}.$$

Proof: This is a countable adaptation of the proof of Theorem 2.4.3. The reader should check the details in the same manner as for that theorem. A set $\mathfrak{B}$ is a σ-field containing $\mathfrak{A}$ if and only if the following condition is met for each $i \in \mathbb{N}$: $\mathfrak{B}_{X_i} \supseteq \mathfrak{A}_{X_i}$, with $\mathfrak{B}_{X_i}$ being a σ-field in $\mathfrak{P}(X_i)$. ∎

CHAPTER 3

LEBESGUE MEASURE

3.1 THE FINITE INTERVAL $[-N, N)$

Let $X_N = [-N, N) \subset \mathbb{R}$, a left-closed, right-open finite interval. Denote an interval $[a_k, b_k)$ by J_k. We will call a set $E = \bigcup_{k=1}^{n} J_k$ an *elementary set* in X_N, provided that

$$-N \leqslant a_1 \leqslant b_1 \leqslant a_2 \leqslant b_2 \leqslant \ldots \leqslant a_n \leqslant b_n < N,$$

and we will denote by $\mathfrak{E}$ the field of such elementary sets in X_N.

The choice of left-closed, right-open intervals may appear to be very restrictive, but this choice makes $\mathfrak{E}$ easily into a field. And it is not a limitation in the long run, since we will show soon that each singleton $\{x\}$ turns out to be a Borel set in X_N, and thus every interval is a Borel set in X_N. Moreover, we will extend the study of Lebesgue measure to all of $\mathbb{R}$ by means of σ-finiteness.

By Exercise 2.13, we can define a finitely additive measure μ on $\mathfrak{E}$ by the formula

$$\mu(E) = \sum_{i=1}^{n} (b_i - a_i)$$

Measure and Integration: A Concise Introduction to Real Analysis. By Leonard F. Richardson
Copyright © 2009 John Wiley & Sons, Inc.

for each $E \in \mathfrak{E}$. Then $\mu(X_N) = 2N$. Since this is a finite total measure, we will apply the result of Exercise 2.19 to prove in the next theorem the extendability of μ to a countably additive measure on $\mathbb{B}(\mathfrak{E})$. Note that μ is simply the extension to the field generated by intervals J_k of the measure that is Euclidean *length* on each interval J_k.

Theorem 3.1.1 *Let $\mathfrak{E}$ be the field of disjoint unions of finitely many intervals $J_k = [a_k, b_k)$ in $[-N, N)$, as described above. There is a countably additive measure μ^* defined on $\mathbb{B}(\mathfrak{E})$ that coincides with the measure μ on $\mathfrak{E}$, where μ of an interval is its Euclidean length.*

Proof: By the Hopf Extension Theorem, Theorem 2.4.1, it is sufficient to prove that μ is countably additive *on* $\mathfrak{E}$. By Exercise 2.19, it suffices to show that for each decreasing sequence of elementary sets

$$E_1 \supseteq E_2 \supseteq \ldots \supseteq E_n \supseteq \ldots$$

such that $\bigcap_{n=1}^{\infty} E_n = \emptyset$, we have $\lim_{n \to \infty} \mu(E_n) = 0$. If this were false, then there would exist $\delta > 0$ such that $\mu(E_n) \geq \delta$ for all $n \in \mathbb{N}$. For each $n \in \mathbb{N}$, we can pick an elementary set E_n', the closure $\bar{E}_n'$ of which is a compact set[22] $\bar{E}_n' \subset E_n$, and such that

$$\mu(E_n \backslash E_n') < \frac{\delta}{2^{n+1}}.$$

We should note that $\bar{E}_n'$ is not an elementary set, because it is not a union of finitely many left-closed, *right-open* intervals. Thus we take care not to apply μ to $\bar{E}_n'$. It is important to note also that the sets E_n' need not form a decreasing sequence. Although

$$E_n \supseteq E_{n+1} \supset E_{n+1}',$$

the fact that E_n' misses a part of E_n having measure less than $\frac{\delta}{2^{n+1}}$ means that E_n' misses at most that much of E_{n+1}'. The most we can conclude is that

$$\mu\left(E_{n+1}' \backslash E_n'\right) < \frac{\delta}{2^{n+1}}.$$

Also, $\bigcap_{n=1}^{K} E_n' \subset \bigcap_{n=1}^{K} E_n$, and

$$\mu\left(\bigcap_{n=1}^{K} E_n \backslash \bigcap_{n=1}^{K} E_n'\right) < \sum_{n=1}^{K} \frac{\delta}{2^{n+1}} < \frac{\delta}{2}.$$

It follows that for each $K \in \mathbb{N}$ we have

$$\mu\left(\bigcap_{n=1}^{K} E_n'\right) \geq \frac{\delta}{2}, \text{ so that } \bigcap_{n=1}^{K} E_n' \neq \emptyset.$$

---

[22]For each interval $[a, b)$ among the finitely many comprising the elementary set E_n, we use an interval of the form $[a, b - \delta)$ for a suitable, very small $\delta > 0$. Thus we can construct E_n' so that its closure will be closed in both the topology of $\mathbb{R}$ and the relative topology of $[-N, N)$.

Thus

$$\bigcap_{n=1}^{K} \bar{E}'_n \supseteq \bigcap_{n=1}^{K} E'_n \neq \emptyset$$

for all $K \in \mathbb{N}$. Yet we know that

$$\bigcap_{n=1}^{\mathcal{L}} E_n \supseteq \bigcap_{n=1}^{\mathcal{L}} \bar{E}'_n = \emptyset.$$

But each $\bar{E}'_n$ is a closed subset of the compact set $[-N, N]$. By the *finite intersection property* (Exercise 3.1) for compact spaces, this is impossible.

Since the hypotheses of the Hopf Extension Theorem are satisfied, there is a countably additive measure l, the restriction of μ^* to the μ^*-measurable sets in $\mathfrak{P}(X_N)$, that extends the concept of the length of an interval. The measure l is called *Lebesgue measure*, and its domain is the set of all *Lebesgue measurable sets* in X_N.

■

We have constructed a Lebesgue measure for X_N for each $N \in \mathbb{N}$. It is important to note that the Lebesgue measure defined on X_N is the restriction to $[-N, N)$ of the Lebesgue measure defined on $X_{N'}$ for each $N' > N$. This follows from Theorem 2.4.3 concerning the Borel field inherited by a subset and from Exercise 2.21 establishing the uniqueness of the Hopf extension measure for σ-finite measure spaces.

We saw in Exercise 2.20 that the outer measure constructed in the Hopf Extension Theorem (Theorem 2.4.1) is regular. This means that every set $S \in \mathfrak{P}(X)$ is a subset of a Borel set $B \in \mathbb{B}(\mathfrak{A})$ having the same measure as the outer measure of S. We apply these concepts to Lebesgue measure constructed in the present section on the interval $[-N, N)$. The field of elementary sets $\mathfrak{E} \subset \mathbb{B}(\mathfrak{E}) \subset \mathfrak{L}$, where we use the letter $\mathfrak{L}$ to *denote the family of all measurable sets* in the case of Lebesgue measure.

EXERCISE

3.1   A topological space X is called *compact* provided that it has the *Heine-Borel property*: Every open covering of X has a finite subcover.

a) Suppose X is compact and $\mathfrak{F}$ is a family of closed subsets of X such that

$$\bigcap \{F \mid F \in \mathfrak{F}\} = \emptyset.$$

Prove that there is a finite subcollection of $\mathfrak{F}$ with empty intersection.

b) Prove the *finite intersection property* for compact sets: If every intersection of finitely many members of $\mathfrak{F}$ is nonempty, then

$$\bigcap \{F \mid F \in \mathfrak{F}\} \neq \emptyset.$$

3.2 MEASURABLE SETS, BOREL SETS, AND THE REAL LINE

The following theorem is concerned with abstract measures on sets, as constructed using the Hopf Extension Theorem. We should recall Exercise 2.20, which tells us that if a countably additive measure μ is generated by the Hopf Extension Theorem from a finitely additive measure μ on a field $\mathfrak{A}$, then the outer measure μ^* is regular in the following sense: For each set $S \in \mathfrak{P}(X)$, there is a Borel set $B \in \mathbb{B}(\mathfrak{A})$ such that $\mu^*(S) = \mu(B)$.

Theorem 3.2.1 *Let μ be a finite, countably additive measure, constructed from some finitely additive measure on a field in $\mathfrak{A} \subseteq \mathfrak{P}(X)$, using the Hopf Extension Theorem (2.4.1). Then a set S in $\mathfrak{P}(X)$ is measurable[23] if and only if there exist Borel sets B_* and B^* in $\mathbb{B}(\mathfrak{A})$ such that*

$$B_* \subseteq S \subseteq B^* \ \ and \ \ \mu(B_*) = \mu^*(S) = \mu(B^*),$$

where μ^* is the outer measure *defined on $\mathfrak{P}(X)$, as in the proof of the Hopf Extension Theorem.*

Proof: We begin with necessity. Suppose that S is measurable. Then we know that $\mu(S) + \mu(S^c) = \mu(X)$. Because μ is regular, there exists a Borel set $B^* \supseteq S$ such that $\mu(S) = \mu(B^*)$. We know that the measurable sets form a σ-field, so S^c is measurable too. So there exists a Borel set $B \supseteq S^c$ such that $\mu(S^c) = \mu(B)$. If we let $B_* = B^c$, then $B_* \subseteq S$ and $\mu(B_*) = \mu(S)$.

Next, we prove sufficiency. Suppose there exist Borel sets $B_* \subseteq S \subseteq B^*$ such that $\mu(B_*) = \mu^*(S) = \mu(B^*)$. We know that S will be measurable provided that $\mu^*(S) + \mu^*(S^c) = \mu(X)$.

Since μ^* is subadditive, it suffices to prove that $\mu^*(S) + \mu^*(S^c) \leqslant \mu(X)$. We know that

$$\mu(B_*) + \mu(B_*^c) = \mu(X) = \mu(B^*) + \mu((B^*)^c),$$

so that $\mu((B_*)^c) = \mu((B^*)^c)$. By the monotonicity of μ^* we see that

$$\mu^*(S) = \mu(B^*) \ \ and \ \ \mu^*(S^c) \leqslant \mu((B_*)^c).$$

Thus

$$\mu^*(S) + \mu^*(S^c) \leqslant \mu(B^*) + \mu((B_*)^c)$$
$$= \mu(B^*) + \mu((B^*)^c) = \mu(X).$$

■

Corollary 3.2.1 *Let μ be a σ-finite, countably additive measure, constructed from some finitely additive measure on a field in $\mathfrak{A} \subseteq \mathfrak{P}(X)$, using the Hopf Extension*

---

[23]The term *μ-measurable* would be used if there were any ambiguity regarding the measure under consideration.

Theorem (2.4.1). Then a set S in $\mathfrak{P}(X)$ *is measurable if and only if there exist Borel sets* B_* *and* B^* *in* $\mathbb{B}(\mathfrak{A})$ *such that*

$$B_* \subseteq S \subseteq B^* \text{ and } \mu(B^* \backslash B_*) = 0.$$

Proof: Apply σ-finiteness, taking note of the fact that the union of countably many sets of μ-measure zero will have μ-measure zero.   ∎

The reader should note that if S is a measurable set, then $\mu^*(S) = \mu(S)$. That is, if $S \in \mathfrak{L}$, then there exist B^* and $B_* \in \mathbb{B} \subset \mathfrak{L}$ such that

$$\mu(B_*) = \mu(S) = \mu(B^*).$$

This is the first result of several that will help us to explore the nature of Lebesgue measurable sets. The exercises below provide very important additional insights that will be useful frequently.

Definition 3.2.1 A set $B \subset \mathbb{R}$ is called an F_σ-*set* if and only if it can be expressed as the union of countably many closed sets. A set $B \subset \mathbb{R}$ is called a G_δ-*set* if and only if it can be expressed as the intersection of countably many open sets. [24]

In view of Exercise 3.2 below, F_σ-sets and G_δ-sets are special types of Borel sets in the real line.

EXERCISES

3.2   Prove that every open set $G \subseteq [-N, N)$ is a Borel set and that every closed set $F \subset [-N, N)$ is a Borel set. In particular, each singleton set $\{p\}$ is a Borel set, and each interval is a Borel set. (Hint: You may use the fact from advanced calculus that every open subset of $\mathbb{R}$ is the union of *countably many* open intervals.)

3.3   Suppose that $S \subseteq [-N, N)$. Use Theorem 3.2.1 to prove that S is a measurable subset of $[-N, N)$ if and only if for each $\epsilon > 0$ there exist an open set $G \supseteq S$ and a closed set $F \subseteq S$ such that $\mu(G) - \mu(F) < \epsilon$. (Hint: The measure of a Borel set is the same as its outer measure. To produce G, think about expanding the half-closed, half-open intervals very slightly to make them open.)

3.4   Prove that in Theorem 3.2.1 we can take the set B_* to be an F_σ-set and the set B^* to be a G_δ-set. (Hint: Use Exercise 3.3.)

Remark 3.2.1   The measure μ^* on the power set of $X_N = [-N, N)$ that we have defined in this section is called *Lebesgue outer measure* and is denoted also as l^*. We define the *Lebesgue inner measure* by the formula

$$l_*(S) = l(X_N) - l^*(S^c).$$

---

[24]These notations are universal in the mathematical literature. The name G_δ comes from the German word *gebiet*, meaning a domain, which is normally open, and from the Greek δ, analogous to the Roman d, for the German word *durchschnitt*, meaning intersection. The name F_σ comes from the French word *fermé*, meaning closed, and the Greek letter σ, analogous to the Roman s, for sum.

It follows from Theorem 2.4.2 that $S \subseteq X_N$ is Lebesgue measurable if and only if $l^*(S) = l_*(S)$, in which case $l(S) = l^*(S) = l_*(S)$.

Definition 3.2.2 We will call the measure l defined in Remark 3.2.1 *Lebesgue measure* on $X_N = [-N, N)$. The *Lebesgue measurable* sets in X_N are the sets in the family

$$\mathfrak{L}(X_N) = \{A \in \mathfrak{P}(X) \mid l^*(W) = l^*(W \cap A) + l^*(W \cap A^c) \; \forall W \in \mathfrak{P}(X)\}.$$

3.2.1 Lebesgue Measure on $\mathbb{R}$

Next, we extend the definitions of Borel sets, measurable sets, and Lebesgue measure to $\mathbb{R}$ from $X_N = [-N, N)$. We will apply Corollary 2.4.1 so as to make use of the σ-finiteness of $\mathbb{R}$ with respect to Lebesgue measure. One may replace the decomposition $\mathbb{R} = \bigcup_{N \in \mathbb{N}} X_N$ by

$$\mathbb{R} = \bigcup_{N \in \mathbb{N}} S_N,$$

where $S_N = X_N \backslash X_{N-1}$ is a sequence of mutually disjoint Borel sets of finite Lebesgue measure.

Definition 3.2.3 We say that $A \in \mathfrak{B}(\mathbb{R})$, the family of all Borel sets in the real line, if and only if $A \cap S_N \in \mathfrak{B}(S_N)$ for each $N \in \mathbb{N}$. Call $A \in \mathfrak{L}(\mathbb{R})$, the family of all measurable sets in the real line, if and only if $A \cap S_N \in \mathfrak{L}(S_N)$ for each $N \in \mathbb{N}$. Finally, define the Lebesgue measure of $A \in \mathfrak{L}(\mathbb{R})$ by

$$l(A) = \sum_{N \in \mathbb{N}} l(A \cap S_N).$$

Theorem 3.2.2 *Let $\mathcal{O}$ denote the set of all open subsets of $\mathbb{R}$. Then the family $\mathfrak{B}$ of all Borel sets in the line is generated by $\mathcal{O}$:*

$$\mathfrak{B}(\mathbb{R}) = \mathbb{B}(\mathcal{O}).$$

Proof: The reader will provide a proof in Exercise 3.5.   ∎

Remark 3.2.2 We began the study of Borel sets and of measurable sets in the real line by considering intervals of the form $[a, b)$, because we need Lebesgue measure to generalize the concept of the length of an interval. The use of left-closed, half-open intervals was a convenience for describing a *field* of sets so that we could use the theorem that the monotone class and Borel field generated by the elementary sets would be the same.

At this point, it is advantageous to think of the σ-field of Borel sets in the real line as being generated by its topology—that is, by the family of open sets. First, this establishes that the family of Borel sets is *canonical*, meaning that it is independent of the choice of decomposition of $\mathbb{R}$ into the union of countably many elementary sets of finite measure. Moreover, this interpretation generalizes to topological spaces that are different from $\mathbb{R}$.

EXERCISES

3.5   Prove each of the following statements about sets $A \subseteq \mathbb{R}$.

   a) Let $\mathfrak{E}_N$ be the field of elementary sets in $S_N = X_N \backslash X_{N-1}$ and denote

$$\mathfrak{E}(\mathbb{R}) = \left\{ E = \bigcup_{n=1}^{N} E_n \;\middle|\; E_n \in \mathfrak{E}_n \; \forall n \leqslant N, \; \forall N \in \mathbb{N} \right\}.$$

   Prove that $\mathfrak{B}(\mathbb{R})$ is the σ-field generated by the field $\mathfrak{E}(\mathbb{R})$. (Hint: Use Corollary 2.4.1.)

   b) Prove Theorem 3.2.2: Show that $\mathfrak{B}(\mathbb{R})$ is the σ-field generated by the family of all open sets in $\mathbb{R}$, or equivalently, generated by the family of all closed sets.

   c) Prove that $A \in \mathfrak{L}(\mathbb{R})$ if and only if there exist B^* and B_* in $\mathfrak{B}(\mathbb{R})$ such that $B_* \subseteq A \subseteq B^*$ and $l(B^* \backslash B_*) = 0$. Explain why it would be insufficient on $\mathbb{R}$ to know only that $\mu(B^*) = \mu(A) = \mu(B_*)$.

   d) Show that l is a countably additive measure on both $\mathfrak{B}(\mathbb{R})$ and $\mathfrak{L}(\mathbb{R})$. (Hint: You may use theorems from advanced calculus concerning absolutely convergent double series.)

3.6   Let E be any measurable subset of $\mathbb{R}$. Define the set

$$E + c = \{e + c \mid e \in E\}.$$

We will prove the *translation invariance* of Lebesgue measure.

   a) Suppose first that E is an open set, and prove that $l(E) = l(E + c)$. (Hint: You may use the fact that every open subset of the real line is the union of countably many *disjoint* open intervals.)

   b) Prove that translation preserves the family of G_δ-sets, the family of F_σ-sets, and the family of Lebesgue null sets.

   c) Prove that E is measurable if and only if $E + c$ is measurable.

   d) Prove that $l(E) = l(E + c)$ for each Lebesgue measurable set E.

   e) *Translations modulo 1:* Now let $T_c : [0, 1) \rightarrow [0, 1)$ by

$$T_c(x) = x + c - \lfloor x + c \rfloor,$$

   where the *floor function* $\lfloor x \rfloor$ denotes the greatest integer that does not exceed x. Prove that T_c preserves both the measurability and the measure of each measurable subset of $[0, 1)$. (Hint: T_n is the identity map if $n \in \mathbb{Z}$. Partition E into two suitable subsets.)

3.7   Let $S = (\mathbb{R}\backslash\mathbb{Q}) \cap [0,1]$, and let l denote Lebesgue measure.
   a) Show that $l(S) = 1$.
   b) Let $\epsilon > 0$. Construct a closed subset $F \subset S$ such $l(F) > 1 - \epsilon$.
   c) Explain why it is impossible for F to contain any interval of length greater than zero.

3.8   Let $S \in \mathfrak{L}$ be a Lebesgue measurable subset of $\mathbb{R}$. Suppose that for each finite interval $I \subseteq \mathbb{R}$ we have $l(S \cap I) \leqslant \frac{1}{2}l(I)$. Prove that $l(S) = 0$.

3.9   Let $E \in \mathfrak{L}(\mathbb{R})$ be any Lebesgue measurable subset of the real line for which $l(E) < \infty$. Let $f(x) = l(E \cap (-\infty, x))$.
   a) Prove that $f \in C(\mathbb{R})$, the vector space of continuous real-valued functions on the real line.
   b) Let $\alpha \in (0,1)$. Prove that there exists a *measurable* set $E_\alpha \subset E$ such that $l(E_\alpha) = \alpha l(E)$.

3.3   MEASURE SPACES AND COMPLETIONS

Definition 3.3.1 A *measure space* is *any* triplet $(X, \mathfrak{A}, \mu)$ where X is a set, $\mathfrak{A}$ is a σ-field of subsets of X, and μ is a countably additive measure on $\mathfrak{A}$. In a measure space, a set $A \in \mathfrak{A}$ is called a *null set* provided that $\mu(A) = 0$. A measure space is called *complete* provided that every subset of a null set belongs to $\mathfrak{A}$. A pair $(X, \mathfrak{A})$ is called a *measurable space* provided that $\mathfrak{A}$ is a σ-field in $\mathfrak{P}(X)$.

EXERCISE

3.10   Let $X = \mathbb{R}$, the real line, and $\mathfrak{A} = \mathfrak{P}(\mathbb{R})$, the σ-algebra of all subsets of X. Define $\nu(E) = \#(E \cap \mathbb{Z})$, the number of integers in E.
   a) Prove that ν is a countably additive measure on $\mathfrak{A}$.
   b) Let $f(x) = \nu([0,x))$. Is f continuous? Justify your answer.
   c) Is $(X, \mathfrak{A}, \nu)$ complete? Justify your answer.

If A is a null set and if $B \in \mathfrak{A}$ is a subset of A, then the monotonicity of μ implies that $\mu(B) = 0$, so that B is a null set. The reason that not every measure space is complete is that a subset of a null set may fail to belong to $\mathfrak{A}$. However, we do have the following corollary of the Hopf Extension Theorem.

Corollary 3.3.1 *Let* $(X, \mathfrak{A}^*, \mu^*)$ *be any measure space produced by applying the Hopf Extension Theorem (Theorem 2.4.1) to produce a Carathéodory outer measure* μ^*, *where* $\mathfrak{A}^*$ *is the σ-field of all μ^*-measurable sets. Then* $(X, \mathfrak{A}^*, \mu^*)$ *is a complete measure space. In particular, each of the Lebesgue measure spaces*

$$([-N, N), \mathfrak{L}[-N, N), l)$$

is complete, as is $(\mathbb{R}, \mathfrak{L}(\mathbb{R}), l)$.

Proof: We will not suppose that $\mu^*(X) < \infty$ since this is not needed. Suppose that $A \in \mathfrak{A}^*$ is a null set, and let $B \subset A$. We need to prove that B is measurable. Thus we need to show that for each $W \in \mathfrak{P}(X)$ we have

$$\mu^*(W) = \mu^*(W \cap B) + \mu^*(W \cap B^c).$$

Since A is measurable, we know that

$$\mu^*(W) = \mu^*(W \cap A) + \mu^*(W \cap A^c),$$

and

$$0 \leqslant \mu^*(W \cap B) \leqslant \mu^*(W \cap A) \leqslant \mu^*(A) = 0.$$

Hence $\mu^*(W \cap B) = \mu^*(W \cap A) = 0$. It follows that

$$\mu^*(W) = \mu^*(W \cap A^c) \leqslant \mu^*(W \cap B^c) \leqslant \mu^*(W)$$

by monotonicity. Thus

$$\mu^*(W) = \mu^*(W \cap A^c) = \mu^*(W \cap B^c) = \mu^*(W \cap B) + \mu^*(W \cap B^c).$$

Hence $B \in \mathfrak{A}^*$.

The real cases $\big([-N, N), \mathfrak{L}[-N, N), l\big)$ are covered directly by the preceding general argument. And the case of $(\mathbb{R}, \mathfrak{L}(\mathbb{R}), l)$ is covered by the completeness of l on $[-N, N)$ for each $N \in \mathbb{N}$, since Lebesgue measure on the real line is σ-finite. ∎

It will follow from Exercise 3.11 that the family of Lebesgue measurable subsets of $[0, 1)$ has cardinality 2^c which is strictly greater than $c = 2^{\aleph_0}$, the cardinality of the set $\mathbb{R}$ of all real numbers. The exercise asks us to construct a Borel set that is known as the *middle thirds Cantor set, C.*

EXERCISE

3.11 Let C_1 be the set $[0, 1] \backslash \left(\frac{1}{3}, \frac{2}{3}\right)$, a union of two disjoint closed intervals, each of length $\frac{1}{3}$. Let

$$C_2 = C_1 \backslash \left(\left(\frac{1}{9}, \frac{2}{9}\right) \cup \left(\frac{7}{9}, \frac{8}{9}\right)\right),$$

so that again we discard the middle thirds, leaving four disjoint closed intervals, each of length $\frac{1}{9}$. We proceed inductively in this manner, each time discarding middle thirds, producing a descending chain of nested sets $C_1 \supset C_2 \supset C_3 \supset \dots$. Define the *Cantor set*

$$C = \bigcap_{n=1}^{\infty} C_n.$$

a) Prove that C is a Borel set with measure zero.

b) Use *ternary expansions* of the real numbers to produce a bijection between the elements of C and the set of all infinite sequences of 0s and 2s. Explain why this set is uncountable and has the cardinality c of the continuum of real numbers.

c) Prove that C is a closed, nowhere dense subset of the unit interval $[0, 1]$.

Remark 3.3.1 We sketch here a proof that the *Borel* measure space

$$(X_N, \mathfrak{B}, \mu),$$

consisting of all the Borel sets in the interval $X_N = [-N, N)$, is *not* complete.[25] The argument will not be self-contained, since it will depend upon outside knowledge of transfinite induction and how that process can be used to construct the σ-algebra $\mathfrak{B}$. What we will do is to show that there are not enough Borel sets to account for all the subsets of the middle thirds Cantor set. We will explain why the cardinality of the set $\mathfrak{B}$ of all Borel sets in X_N has the cardinality

$$c = 2^{\aleph_0} < 2^c = \#\mathfrak{P}(C),$$

where C is the middle thirds Cantor set.

The interested reader can consult [8] for the details of the following construction by transfinite induction of $\mathfrak{B}$. For each set $S \subseteq \mathfrak{P}(X)$, denote by S^* the set of all unions of countably many members of S and all differences of elements of S. Let $k = \#S$, the (generally infinite) cardinality of S. Sequences in S correspond to functions $f : \mathbb{N} \to S$, and the cardinality of the set of these is the (infinite) cardinal number $k^{\aleph_0}$. Suppose for the moment that $k = c = 2^{\aleph_0}$. Then

$$k^{\aleph_0} = c^{\aleph_0} = \left(2^{\aleph_0}\right)^{\aleph_0} = 2^{(\aleph_0 \times \aleph_0)} = 2^{\aleph_0} = c,$$

because for any infinite cardinal numbers[26] $\beth$ and $\gimel$ we have

$$\beth \times \gimel = \max(\beth, \gimel).$$

Thus the cardinality

$$\#S^* = c = \#S.$$

First, let $\mathfrak{E} \subset \mathfrak{P}(X)$ be the field of elementary sets. It is easy to see that the cardinality of $\mathfrak{E}$ is c. Let $\mathfrak{E}_0 = \mathfrak{E}$, and for each finite or transfinite ordinal number α let

$$\mathfrak{E}_\alpha = \left(\bigcup_{\beta < \alpha} E_\beta\right)^*.$$

---

[25] A different proof, based on the Cantor function, of the existence of measurable sets that are not Borel sets is given in Exercise 7.13.c.

[26] The smallest infinite cardinal number is the cardinality of the set $\mathbb{N}$ of natural numbers and it is written as $\aleph_0$ (read *aleph-null*). The next two symbols we have used are *bet* $\beth$ and *gimel* $\gimel$, which are the next two letters of the Hebrew alphabet, or *aleph-bet*. Each infinite cardinal number is an equivalence class of infinite sets. Each infinite ordinal number is an order-preserving equivalence class of well-ordered sets.

Let Ω denote the first uncountable ordinal number. It can be shown that

$$\mathbb{B}(\mathfrak{E}) = \bigcup_{\alpha < \Omega} E_\alpha$$

and that the cardinality of $\mathbb{B}(\mathfrak{E})$ is $c \times c = c$, the product of any two infinite cardinal numbers being the larger of the two.

3.3.1   Minimal Completion of a Measure Space

Theorem 3.3.1 *For each measure space* $(X, \mathfrak{A}, \mu)$ *there exists a* minimal completion $(X, \bar{\mathfrak{A}}, \bar{\mu})$. *That is,* $(X, \bar{\mathfrak{A}}, \bar{\mu})$ *is a complete measure space that is an* extension *of* $(X, \mathfrak{A}, \mu)$ *in the sense that* $\bar{\mathfrak{A}} \supseteq \mathfrak{A}$ *and the* restriction

$$\bar{\mu}\big|_{\mathfrak{A}} = \mu.$$

Moreover, every complete extension of $(X, \mathfrak{A}, \mu)$ *is an extension of* $(X, \bar{\mathfrak{A}}, \bar{\mu})$.

Proof. Let $(X, \mathfrak{A}, \mu)$ be any measure space. Let

$$\mathfrak{N} = \{A \mid A \in \mathfrak{A}, \mu(A) = 0\},$$
$$\bar{\mathfrak{N}} = \{B \mid B \subseteq A \text{ for some } A \in \mathfrak{N}\},$$
$$\bar{\mathfrak{A}} = \{\bar{A} \mid \bar{A} = (A \backslash N_1) \cup N_2, A \in \mathfrak{A}, N_1 \in \bar{\mathfrak{N}}, N_2 \in \bar{\mathfrak{N}}\},$$

and let $\bar{\mu}\big((A \backslash N_1) \cup N_2\big) = \mu(A)$. It is easy to see that $\bar{\mathfrak{A}}$ includes every subset of every null set, so that $(X, \bar{\mathfrak{A}}, \bar{\mu})$ is complete. The reader should check that $\bar{\mathfrak{A}}$ is a σ-field. This follows from the fact that countable unions of null sets are null sets. Also, every complete extension of $(X, \mathfrak{A}, \mu)$ must extend $(X, \bar{\mathfrak{A}}, \bar{\mu})$, so that the latter completion is minimal. ∎

Corollary 3.3.2 *A σ-finite measure space* $(X, \mathfrak{L}, \mu)$, *arising from the Hopf Extension Theorem using the σ-field $\mathfrak{L}$ of all the measurable sets, is the (minimal) completion of the Borel measure space* $(X, \mathfrak{B}, \mu)$. *(Here, $\mathfrak{B}$ is the Borel field generated by the original field in the hypothesis of the Hopf theorem.)*

Proof: We have seen that $(X, \mathfrak{L}, \mu)$ is complete by Corollary 3.3.1. By Corollary 3.2.1, every measurable set A in a σ-finite measure space lies between two Borel sets as follows: $B_* \subseteq A \subseteq B^*$, where $\mu(B^* \backslash B_*) = 0$. Thus

$$A = (B_* \cup N_1) \backslash N_2,$$

where N_1 and N_2 are subsets of the null set $B^* \backslash B_*$). Hence each $A \in \mathfrak{L}$ must lie in the minimal completion of $(X, \mathfrak{B}, l)$, as defined in the proof of Theorem 3.3.1. ∎

3.3.2   A Nonmeasurable Set

We prove here the existence of a nonmeasurable set for Lebesgue measure on the unit interval.

■ **EXAMPLE 3.1**

We will prove the existence of a *nonmeasurable* set, specifically within the unit interval of the real line.[27] Define a translation mapping of $X = [0, 1)$ onto itself by

$$T_c(x) = x + c - \lfloor x + c \rfloor,$$

where $\lfloor x + c \rfloor$ denotes the *greatest integer* that does not exceed x. This can be understood as *translation modulo one*. It follows from Exercise 3.6 that if $E \subseteq X$ is measurable, then $T_c(E)$ is measurable as well, and $l(E) = l(T_c(E))$. Now define an equivalence relation on X by $x \sim y$ if and only if $x - y \in \mathbb{Q}$, the set of rational numbers. Decompose X into disjoint equivalence classes:

$$ \cdots = \bigcup_{\gamma \in \Gamma} A_\gamma. $$

By the Axiom of Choice there exists a set

$$M = \{x_\gamma \mid x_\gamma \in A_\gamma \ \forall \gamma \in \Gamma\}$$

consisting of exactly one member x_γ from each of the equivalence classes A_γ. We claim that $M \notin \mathfrak{L}$. Note that $x_\gamma - x_{\gamma'} \in \mathbb{Q}$ if and only if $\gamma = \gamma'$. Hence the family of sets

$$\{T_c(M) \mid c \in \mathbb{Q} \cap [0, 1)\}$$

is a countable family of mutually disjoint sets. Moreover,

$$X = \dot{\bigcup}_{c \in \mathbb{Q} \cap X} T_c(M).$$

If M were measurable, then either

$$l(M) = 0$$

or

$$l(M) > 0.$$

In the former case, countable additivity of Lebesgue measure would imply that $l(X) = 0$, which is impossible since $l(X) = 1$. But if $l(M) > 0$, it would follow that $l(X) = \infty$, which is impossible as well. Hence $M \notin \mathfrak{L}$.   ■

EXERCISE

3.12   Let S be any nonmeasurable subset of $\mathbb{R}$. Prove that the *outer* measure $l^*(S) > 0$.

---

[27]A more general result is presented soon as Theorem 3.4.4, the proof of which utilizes Steinhaus's theorem from Exercise 3.21. A further development of the construction of a nonmeasurable set appears in the Appendix as Example A.1.

3.4   SEMIMETRIC SPACE OF MEASURABLE SETS

Definition 3.4.1 A set S equipped with a function $d : S \times S \to [0, \infty)$ is called a *metric space* (S, d), with *metric* d, provided that d has the following properties:

  i. $d(a, b) = d(b, a)$ for all a and b in S.

  ii. $d(a, b) = 0$ if and only if $a = b$.

  iii. $d(a, c) \leqslant d(a, b) + d(b, c)$ for all a, b, and c in S.

Property (iii) is called the *triangle inequality*. If the pair (S, d) lacks the *only if* part of property (ii) but has the others, then it is called a *semimetric space*. In any metric or semimetric space, a sequence a_n is called a *Cauchy sequence* if and only if for each $\epsilon > 0$ there exists $N \in \mathbb{N}$ such that n and m greater than or equal to N implies that $d(a_n, a_m) < \epsilon$. A metric or semimetric space is called *complete* if and only if for each Cauchy sequence a_n in S there exists $a \in S$ such that $a_n \to a$ in the sense that $d(a_n, a) \to 0$ as $n \to \infty$.[28]

Theorem 3.4.1 Let $(X, \mathfrak{A}, \mu)$ be any finite measure space. For each pair of sets A and B in $\mathfrak{A}$, define the symmetric difference $A \triangle B = (A \cup B) \setminus (A \cap B)$ and define $d(A, B) = \mu(A \triangle B)$. Then the pair $(\mathfrak{A}, d)$ is a complete semimetric space.

The reader should note that this concept of completeness *as a metric space* does not have the same meaning as completeness of a measure space in the previous sense, in which every subset of a null set is a null set. The present theorem does not require the *measure space* to be complete in the sense of Theorem 3.3.1.

Proof: To prove that d is a semimetric, everything except the triangle inequality is obvious. To establish the triangle inequality, it would suffice to prove what we will call the *triangle inequality for sets:*

$$A \triangle C \subseteq (A \triangle B) \cup (B \triangle C) \tag{3.1}$$

To prove this containment, we note first that if $x \in A \triangle C$, then either $x \in A$ or $x \in C$, but not both. Thus $A \triangle C = (A \setminus C) \cup (C \setminus A)$. Hence if $x \in A \triangle C$, it follows that either

$$x \in A \setminus C, \text{ and thus } \begin{cases} x \in B, \text{ and } x \in B \triangle C, \text{ or} \\ x \notin B, \text{ and } x \in A \triangle B, \end{cases}$$

or else

$$x \in C \setminus A, \text{ and thus } \begin{cases} x \in B, \text{ and } x \in A \triangle B, \text{ or} \\ x \notin B, \text{ and } x \in B \triangle C. \end{cases}$$

---

[28] However, for limits to be unique, we need the space to be a full-fledged metric space.

In order to prove that the semimetric space is complete, we take a Cauchy sequence $\{A_n \mid A_n \in \mathfrak{A} \; \forall n \in \mathbb{N}\}$. That is, for each $\epsilon > 0$ there exists $N \in \mathbb{N}$ such that n and $m \geqslant N$ implies that $d(A_n, A_m) < \epsilon$. We need to prove that there exists $A \in \mathfrak{A}$ such that $d(A_n, A) \to 0$ as $n \to \infty$.

For each $k \in \mathbb{N}$ there exists $n_k \in \mathbb{N}$ such that n and $m \geqslant n_k$ implies that

$$d(A_n, A_m) < \frac{1}{2^k}.$$

Moreover, we can select the sequence n_k so that $n_k < n_{k+1}$ for all k. We will define the *limit superior* of the sequence of sets A_{n_k} by

$$A = \limsup_{k \to \infty} A_{n_k} = \bigcap_{p=1}^{\infty} \bigcup_{k=p}^{\infty} A_{n_k},$$

the set of all elements that are present in infinitely many of the sets A_{n_k}. Observe that $A \in \mathfrak{A}$ because $\mathfrak{A}$ is a σ-field. We will prove that $d(A_n, A) \to 0$ as $n \to \infty$.

Denote

$$C_p = \bigcup_{k=p}^{\infty} A_{n_k} \text{ and } D_p = \bigcap_{k=p}^{\infty} A_{n_k}.$$

Observe that C_p is a descending sequence of sets with intersection equal to A. On the other hand, D_p is an ascending sequence of sets with union denoted as

$$\liminf A_{n_k} = \bigcup_{p=1}^{\infty} \bigcap_{k=p}^{\infty} A_{n_k}.$$

The *limit inferior* of the sequence of sets A_{n_k} is therefore the set of all elements that are present in all but finitely many of the sets A_{n_k}. Observe that

$$D_p \subseteq A_{n_p} \subseteq C_p, \text{ and } D_p \subseteq A \subseteq C_p,$$

for all p. Thus

$$A \bigtriangleup A_{n_p} \subseteq C_p \bigtriangleup D_p,$$

so that $d(A_{n_p}, A) \leqslant d(A_{n_p}, C_p) + d(C_p, D_p)$. Furthermore,

$$d(C_p, D_p) = \mu(C_p \bigtriangleup D_p)$$

$$= \mu\left(\bigcup_{k=p}^{\infty} A_{n_k} \setminus \bigcap_{k=p}^{\infty} A_{n_k} \right)$$

$$\overset{(i)}{\leqslant} \mu\left(\bigcup_{k=p}^{\infty} \left(A_{n_k} \bigtriangleup A_{n_{k+1}} \right) \right)$$

$$\leqslant \sum_{k=p}^{\infty} \frac{1}{2^k} = \frac{1}{2^{p-1}} \to 0$$

as $p \to \infty$. The inequality (i) is justified by the fact that in order for an element x to be in the union of the pth *tail* of the sequence of sets but not in its intersection, there must be at least one of the sets A_{n_k} in the union that contains x and another set A_{n_j} that does not contain x. Iterated application of Equation (3.1) using in place of the set B each of the sets indexed between n_k and n_j shows that there must be at least one value of k such that this element lies in $A_{n_k} \triangle A_{n_{k+1}}$. In fact, suppose for specificity that $n_j < n_k$. Then

$$A_{n_j} \triangle A_{n_k} \subseteq \bigcup_{i=j}^{k-1} \left(A_{n_i} \triangle A_{n_{i+1}} \right).$$

Thus

$$
\begin{aligned}
d(A_{n_p}, A) &\leqslant d(A_{n_p}, C_p) + d(C_p, A) \\
&\leqslant \mu(C_p \setminus A_{n_p}) + \mu(C_p \setminus A) \\
&\leqslant 2\mu(C_p \setminus D_p) \to 0
\end{aligned}
$$

as $p \to \infty$. Hence if $n > n_p$ we have

$$
\begin{aligned}
d(A_n, A) &\leqslant d(A_n, A_{n_p}) + d(A_{n_p}, A) \\
&< \frac{1}{2^p} + \frac{1}{2^{p-1}} \to 0
\end{aligned}
$$

as $p \to \infty$. ∎

Remark 3.4.1 We could replace the semimetric space of measurable sets in Theorem 3.4.1 by a metric space in which the points of the space are equivalence classes. Define $A \sim B$ if and only if $\mu(A \triangle B) = 0$. That is, two sets A and B are equivalent if and only if the measure of their symmetric difference is zero. We form the quotient space $\mathfrak{A}/\mathfrak{N}$ where $\mathfrak{N}$ denotes the family of null sets in $\mathfrak{A}$. And if $\bar{A}$ and $\bar{B}$ are equivalence classes, we define the full-fledged metric $\bar{d}(\bar{A}, \bar{B}) = d(A, B)$. The reader should verify that the value of the metric is independent of which representatives are selected from each of the two equivalence classes.

In the next theorem, we apply what we have learned about the metric on the space of measurable sets to Lebesgue measure on $X_N = [-N, N)$.

Theorem 3.4.2 *In the semimetric space* $(\mathfrak{A}, d)$ *obtained from the Borel measure space* $([-N, N), \mathfrak{A}, l)$ *by letting* $d(A, B) = l(A \triangle B)$, *the field* $\mathfrak{E}$ *of elementary sets is a* dense *subset of* $\mathfrak{A}$. *That is, if* $B \in \mathfrak{A}$, *then there exists a sequence of elementary sets* $E_n \in \mathfrak{E}$ *such that* $E_n \to B$ *in the sense that* $d(B, E_n) \to 0$ *as* $n \to \infty$.

The proof is Exercise 3.13.

EXERCISES

3.13   Prove Theorem 3.4.2. (Hint: Use the definition of Carathéodory outer measure to show that there exists a suitable sequence of elementary sets.)

3.14   In the semimetric space $(\mathfrak{L}, d)$ formed from $([-N, N), \mathfrak{L}, l)$, where l is Lebesgue measure, prove that each of the following sets is dense:
   a) $\mathcal{O}$, the set of all *open* sets in $[-N, N)$.
   b) $\mathcal{K}$, the set of all *closed* sets in $[-N, N)$.

3.15   Let $(X, \mathfrak{A}, \mu)$ be any finite measure space arising from the Hopf Extension Theorem applied to a finitely additive measure on some field $\mathfrak{E}$.
   a) Prove that Theorem 3.4.2 is true for $(X, \mathfrak{A}, \mu)$.
   b) Explain why we need the measure $\mu(X) < \infty$ in part (a).

The following concept is purely topological and is not part of measure theory, although it can be applied to metric spaces that arise from measure theory as explained above.

Definition 3.4.2 If (X, d) is *any* metric space, a subset $S \subseteq X$ is said to be a set of the *first category* if and only if S can be written as the union of countably many *nowhere dense* sets.[29] A subset $S \subseteq X$ is said to be of the *second category* if and only if it is not of the first category.

For example, the set $\mathbb{Q}$ is a set of the first category in $\mathbb{R}$, equipped with its usual Euclidean topology, since

$$\mathbb{Q} = \bigcup_{q \in \mathbb{Q}} \{q\},$$

a countable union of singleton sets. And each singleton is clearly closed and nowhere dense.

Theorem 3.4.3 (Baire Category Theorem) *A complete metric space is a set of the second category.*

The proof is given in Exercise 3.16.

In the Baire Category theorem the concepts of being closed, dense, or nowhere dense are understood in terms of the given metric. The reader can apply the Baire Category theorem to the complete metric space $\mathfrak{A}/\mathfrak{N}$ of equivalence classes of Lebesgue measurable sets of the measure space $([-N, N), \mathfrak{L}, l)$ in order to prove that measurable sets with a certain bizarre property are ubiquitous. See Exercise 3.17. The Baire Category theorem has many other interesting consequences, some of which are treated in the exercises below.

---

[29]*Nowhere dense* means that the closure $\bar{S}$ of S contains no open set other than the empty set.

EXERCISES

3.16 Prove the Baire Category theorem (3.4.3). Here is a suggested outline.
  a) Show that it suffices to prove that in a complete metric space the intersection of countably many open, dense sets must be nonempty.
  b) Let O_n be a sequence of open, dense subsets of the metric space (X, ρ). Prove that there exists a decreasing nest of closed balls

$$\bar{B}_{r_n}(x_n) \subseteq \bigcap_{k=1}^{n} O_k,$$

  with $r_n \to 0$.
  c) Use completeness to prove that $\bigcap_{k=1}^{\infty} O_k \neq \emptyset$.

3.17 Call the Lebesgue measurable set $S \subset [-N, N)$ a *ghost set* if and only if S has the property that for *every interval* $I \subseteq [-N, N)$ we have

$$0 < l(S \cap I) < l(I).^{30}$$

  a) Show that the family $\mathfrak{G}$ of ghost sets in the unit interval is a set of the second category.
  b) Determine whether $\mathfrak{G}$ is closed under
      i. unions of two elements.
      ii. complementation of an element.
      iii. intersection of two elements.
  c) Let G be a ghost set in $[-1, 1)$ and prove that 1_G is *not* Riemann integrable on $[-1, 1]$. (This exercise shows that in the sense of Baire category, most indicator functions of measurable sets in a finite interval are not Riemann integrable. However, the reader will see that they are all Lebesgue integrable.)

3.18
  a) Use the Baire Category theorem[31] to prove that the set $(\mathbb{R}\backslash\mathbb{Q}) \cap [a, b]$ is not an F_σ set if $a < b$.
  b) Show that the set $\mathbb{Q} \cap [a, b]$ is not a G_δ set if $a < b$. (Hint: If $\mathbb{Q}$ and $\mathbb{Q}^c$ were both G_δ-sets, then the empty set would be the intersection of countably many dense open sets, which would violate the Baire Category theorem.)
  c) Give an example of a subset S of the real line that is a Borel set but is neither an F_σ nor a G_δ set.

3.19 Let S be the set of points x at which a function $f : \mathbb{R} \to \mathbb{R}$ is *discontinuous*.

---

[30]The behavior of ghost sets makes an interesting contrast with Exercise 3.8.
[31]In this problem, think of $[a, b]$ as being a complete metric space with respect to ordinary Euclidean distance on the real line.

a) Prove that S must be an F_σ-set. (Hint: It may be helpful to consider the concept of the *oscillation*

$$o_f(p) = \limsup_{x \to p} f(x) - \liminf_{x \to p} f(x),$$

as defined in advanced calculus.[32] You may use the fact that a function f is continuous at p if and only if $o_f(p) = 0$.)

b) Prove that it is impossible for the set S to be precisely the set $\mathbb{Q}^c$ of all irrational numbers. (Use the result of Exercise 3.18.a.)

c) Give an example of a bounded monotone function f for which the set S of all points of discontinuity is precisely $\mathbb{Q}$.[33]

3.20   Equip the space $C(\mathbb{R})$ of all continuous functions on the real line with the standard sup-norm: $\|f\|_{\sup} = \sup\{|f(x)| \mid x \in \mathbb{R}\}$. Prove that for each $n \in \mathbb{N}$ the set

$$S_n = \left\{ f \in C(\mathbb{R}) \,\middle|\, \inf_{\{x \in \mathbb{R} \mid \exists f'(x)\}} |f'(x)| > n \right\}$$

is an open, dense subset of $C(\mathbb{R})$. Apply the Baire Category Theorem to prove that the set of nowhere differentiable continuous functions is a set of the second category in $C(\mathbb{R})$.[34] (Hint: Think in terms of sawtooth functions.)

3.21   Prove the following theorem of *Steinhaus*. Suppose $A \subset \mathbb{R}$ is Lebesgue measurable and suppose its Lebesgue measure $l(A) > 0$. Denote

$$A - A = \{x - y \mid x \in A, \, y \in A\}.$$

Prove that $A - A$ contains an open interval around 0.
Explain why it would suffice to give a proof if $0 < l(A) < \infty$. Let

$$f(x) = l\big((x + A) \cap A\big).$$

It may help to use the following procedure.[35]

a) Show that if $f(x) > 0$, then $x \in A - A$.

b) Show that Steinhaus's theorem follows if f is continuous at 0.

c) Show that

$$|l(A) - l(B)| \leqslant l(A \triangle B) = d(A, B)$$

if A and B are measurable sets of finite measure and if d is the semimetric for the space of measurable sets.

[32] See for example the book [20].
[33] See, for example, Exercise 7.10 in [20].
[34] This problem is not about measure and integration. We include it here because of its general interest as a corollary to the Baire Category Theorem.
[35] An alternative proof of Steinhaus's theorem is given in Exercise 6.10.

d) Show that f is continuous at each point x if A is a finite interval. Then prove the same thing if A is an elementary set. (It will be easier to give a proof for these two special cases.)

e) Use the triangle inequality for d, together with the result of part (d), to show that f is continuous at each $x \in \mathbb{R}$, assuming only that $A \in \mathcal{L}$ with finite measure.

With the help of Exercise 3.21 we can prove a generalization of Example 3.1. We will show that every subset $P \subseteq \mathbb{R}$ of strictly positive measure must contain a nonmeasurable set.

Theorem 3.4.4 *If $P \in \mathcal{L}(\mathbb{R})$ and if $l(P) > 0$, then there exists a nonmeasurable subset of P.*

Proof: Let $\mathbb{Q}$ denote the set of all rational numbers, as usual. We will define again an equivalence relation on $\mathbb{R}$ by

$$x \sim y \Leftrightarrow x - y \in \mathbb{Q}.$$

By the Axiom of Choice, we can find a cross section Γ [36] of the quotient space $\mathbb{R}/\sim$. We can express the real line as a disjoint union

$$\mathbb{R} = \dot{\bigcup}_{q \in \mathbb{Q}} (\Gamma + q),$$

since if $\gamma + q = \gamma' + q'$, then $\gamma - \gamma' \in \mathbb{Q}$. This would make $\gamma \sim \gamma'$ and thus $\gamma = \gamma'$, since Γ is a cross section of $\mathbb{R}/\sim$.

If $P \cap (\Gamma + q)$ were not Lebesgue measurable for some $q \in \mathbb{Q}$, then we would be finished because P would have a nonmeasurable subset. However, if

$$P \cap (\Gamma + q) = P_q \in \mathcal{L}(\mathbb{R})$$

for each $q \in \mathbb{Q}$, then we observe that

$$P_q - P_q \subseteq (\Gamma + q) - (\Gamma + q) = \Gamma - \Gamma,$$

where $\Gamma - \Gamma$ is disjoint from the dense set $\mathbb{Q}\backslash\{0\}$ for the reasons explained just above. Thus the difference $P_q - P_q$ of a supposedly Lebesgue measurable set with itself fails to include any open interval around 0. By Steinhaus's theorem (Exercise 3.21), P_q must have measure zero. Hence P itself is the union of countably many disjoint null sets, which contradicts the hypothesis that $l(P) > 0$. ∎

More information about the nonmeasurable subsets of P can be found in [10].

EXERCISE

3.22 Prove that the cardinality of the family of all nonmeasurable subsets of $\mathbb{R}$ is the same as the cardinality of the family of all measurable sets: $2^{2^{\aleph_0}}$. (Hint: If P is a nonmeasurable set and S is any disjoint measurable set, what can you conclude about $S \cup P$?)

---

[36]That is, Γ contains exactly one element of each equivalence class in $\mathbb{R}$.

3.5   LEBESGUE MEASURE IN $\mathbb{R}^n$

By the $2N$-*cube* in $\mathbb{R}^n$ we mean

$$Q_N = [-N, N)^n = [-N, N) \times \ldots \times [-N, N),$$

a Cartesian product with n factors, each being a copy of the interval $[-N, N)$. We will denote a typical rectangular block in Q_N by

$$R = J_1 \times \ldots \times J_n,$$

a Cartesian product of n intervals of the form $J_i = [a_i, b_i)$. We define the measure

$$l(R) = \prod_{i=1}^{n} |J_i|,$$

the product of the lengths of the n intervals. By an *elementary set* $E \in \mathfrak{E}$ we will mean the union of finitely many rectangular blocks as defined above, and l is a finitely additive measure on $\mathfrak{E}$.

Theorem 3.5.1 *The family* $\mathfrak{E}(Q_N)$ *of elementary sets is a* field *of sets contained in* $\mathfrak{P}([-N, N)^n)$. *There is a countably additive measure* l *defined on the generated Borel field* $\mathfrak{B} = \mathbb{B}(\mathfrak{E}(Q_N))$ *that coincides with the measure* l *on* $\mathfrak{E}(Q_N)$, *where* l *of a rectangular block is its Euclidean volume.*

The extension of l from $\mathfrak{E}$ to $\mathfrak{B}$ proceeds as it did in Theorem 3.1.1 for the one-dimensional case, using the Hopf Extension Theorem. The proof is virtually identical, since the main tool is the finite intersection property for compact sets. That principle, presented in Exercise 3.1, is valid for $\mathbb{R}^n$ as well. We leave it to the reader to check this fact.

Next we extend the definitions of Borel sets, measurable sets, and Lebesgue measure to $\mathbb{R}^n$ from $Q_N = [-N, N)^n$. We will make use of the σ-finiteness of Euclidean space, just as we did for the line in Definition 3.2.3. One may express

$$\mathbb{R}^n = \overset{\cdot}{\bigcup_{N \in \mathbb{N}}} S_N,$$

where $S_N = Q_N \backslash Q_{N-1}$ for each $N \in \mathbb{N}$. Thus the mutually disjoint elementary sets S_N all have finite Lebesgue measure.

Definition 3.5.1 We say that $A \in \mathfrak{B}(\mathbb{R}^n)$, the family of all Borel sets in $\mathbb{R}^n$, if and only if $A \cap S_N \in \mathfrak{B}(S_N)$ for each $N \in \mathbb{N}$. Call $A \in \mathfrak{L}(\mathbb{R}^n)$, the family of all measurable sets in $\mathbb{R}^n$, if and only if $A \cap S_N \in \mathfrak{L}(S_N)$ for each $N \in \mathbb{N}$. Finally, define the Lebesgue measure of $A \in \mathfrak{L}(\mathbb{R}^n)$ by

$$l(A) = \sum_{N \in \mathbb{N}} l(A \cap S_N).$$

Theorem 3.5.2 *Let $\mathcal{O}$ denote the set of all open subsets of $\mathbb{R}^n$. Then the family $\mathfrak{B}$ of all Borel sets in $\mathbb{R}^n$ is generated by $\mathcal{O}$:*

$$\mathfrak{B}(\mathbb{R}^n) = \mathbb{B}(\mathcal{O}).$$

Proof: This proof is left to the reader in Exercise 3.23.   ∎

Remark 3.5.1 There are advantages to thinking of the σ-field of Borel sets in Euclidean space as being generated by its topology—that is, by the family of open sets. This is a viewpoint that generalizes in interesting ways—for example, to topological groups. We began the study of Borel sets and of measurable sets in the real line by considering half-closed, half-open rectangular blocks, because we require Lebesgue measure to generalize the concept of the volume of a box. The use of half-closed, half-open blocks was primarily a convenience for describing a *field* of sets, so that we could use the theorem that the monotone class and Borel field generated by the elementary sets would be the same.

EXERCISES

3.23 Prove each of the following statements about sets $A \subseteq \mathbb{R}^n$:
   a) Prove that $\mathfrak{B}(\mathbb{R}^n)$ is the σ-field generated by

$$\bigcup_{N \in \mathbb{N}} \mathfrak{E}(S_N).$$

   b) Prove that $A \in \mathfrak{L}(\mathbb{R}^n)$ if and only if there exist B^* and $B_* \in \mathfrak{B}(\mathbb{R}^n)$ such that $B_* \subseteq A \subseteq B^*$ and $l(B^* \backslash B_*) = 0$.
   c) Prove that l is a countably additive measure on $\mathfrak{B}(\mathbb{R}^n)$.
   d) Prove that the open sets and the closed sets are Borel sets in $\mathbb{R}^n$.
   e) Prove Theorem 3.5.2: $\mathfrak{B}(\mathbb{R}^n)$ is the σ-field generated by the family of open sets or, equivalently, by the family of closed sets in $\mathbb{R}^n$.

3.24 Suppose E is Lebesgue measurable in $\mathbb{R}^n$, and define

$$E + c = \{e + c \mid e \in E\},$$

the translate of E by c. (Addition refers to *vector* addition in this context.) We will prove the *translation invariance* of Lebesgue measure.
   a) Prove that E is a null set if and only if $E + c$ is a null set.
   b) Prove that $E + c$ is Lebesgue measurable.
   c) Suppose that E is an *elementary set* in $\mathbb{R}^n$, and prove that $l(E+c) = l(E)$.
   d) Let E be any *measurable set*, and prove that $l(E+c) = l(E)$.

3.25 Suppose E is a subset of $\mathbb{R}^n$ and let $-E = \{-e \mid e \in E\}$. We will prove the *invariance* of Lebesgue measure under *reflection* through the origin.
   a) Suppose first that E is an elementary set in $\mathbb{R}^n$, and prove that $-E$ is a Borel set and $l(-E) = l(E)$.

b) Finally, let E be a measurable set, and prove that $l(-E) = l(E)$.

3.26   Suppose A and B are measurable subsets of $\mathbb{R}^n$, each one of strictly positive but finite measure. Prove that there exists a vector $c \in \mathbb{R}^n$ such that $l((A+c) \cap B) > 0$. (Hint: Consider the outer measure of A and B.)

3.27   Let $\epsilon > 0$. Construct an *open, dense* subset S of $\mathbb{R}^n$ for which the Lebesgue measure $l(S) < \epsilon$.

3.6 JORDAN MEASURE IN $\mathbb{R}^N$

Lebesgue measure will be the foundation for defining the Lebesgue integral and for proving its properties. Jordan measure is the corresponding foundation for the Riemann integral, though Jordan measure (also called *Jordan content*) is often not taught explicitly in advanced calculus courses. Although not required for understanding the Lebesgue integral itself, the study of Jordan measure will help us to understand the relationship between the Lebesgue integral and the Riemann integral, which the Lebesgue integral supersedes. Moreover, Jordan measure will enable us to prove Lebesgue's theorem (6.3.1), classifying all Riemann integrable functions in terms of the Lebesgue measure of the set of points of discontinuity.

Definition 3.6.1 Let $\mathcal{E}$ be the collection of unions of finitely many closed blocks in the closed cube $\bar{Q}_N$, where $N \in \mathbb{N}$ is fixed arbitrarily. Define the *outer Jordan measure* for each $A \in \mathfrak{P}(Q_N)$ by

$$v^*(A) = \inf\{l(E) \mid A \subseteq E, E \in \mathcal{E}\}$$

and let the *inner Jordan measure* be defined by

$$v_*(A) = \sup\{l(E) \mid A \supseteq E, E \in \mathcal{E}\}.$$

Here $l(E)$ denotes the volume, or Lebesgue measure, of the union of finitely many rectangular blocks that comprise E. We define the set $\mathfrak{J}$ to be the family of all Jordan measurable sets, where A is Jordan measurable if and only if $v^*(A) = v_*(A)$, in which case either number is called $v(A)$, the *Jordan measure* of A.

We will see that the weakness of Jordan measure stems from the need to cover a set using only unions of *finitely* many rectangular blocks. This has the unfortunate effect that Jordan measure is only finitely additive, and that is insufficient for the needs of analysis.

EXERCISES

3.28   Prove that Jordan measure is *not* countably additive.

3.29   Use Definition 3.2.1 and DeMorgan's Laws to prove the following properties of inner and outer Jordan measure, in relation to inner and outer Lebesgue measure, for all $A \in \mathfrak{P}(\bar{Q}_N)$. The sets E_i are in $\mathcal{E}$.

a) $l^*(A) = \inf \left\{ \sum_{i \in \mathbb{N}} l(E_i) \,\middle|\, A \subseteq \bigcup_{i \in \mathbb{N}} E_i \right\}.$

b) $v_*(A) \leqslant l_*(A) \leqslant l^*(A) \leqslant v^*(A).$

It follows that every Jordan measurable set is Lebesgue measurable and that its Lebesgue measure equals its Jordan measure. We will see that not every Lebesgue measurable set is Jordan measurable, however. Thus Lebesgue measure is an extension of Jordan measure.

Theorem 3.6.1 *The family* $\mathfrak{J}$ *of all Jordan measurable sets in* $X = \bar{Q}_N$ *is a field, and* v *is a finitely additive measure on* $\mathfrak{J}$.

Proof: Let A and B lie in $\mathfrak{J}$. Since

$$v(X) - v^*(X \backslash A) = v_*(A) \text{ and } v(X) - v_*(X \backslash A) = v^*(A),$$

it follows that $X \backslash A \in \mathfrak{J}$. Both $A \cup B$ and $A \cap B$ are in $\mathfrak{J}$ since the union and intersection of any two elementary sets is again an elementary set. And v is finitely additive on $\mathfrak{J}$ since on that field of sets v agrees with l. ∎

EXERCISE

3.30 A subset A of $X = \bar{Q}_N$ is Jordan measurable if and only if for each $\epsilon > 0$ there exist in $\mathcal{E}$ sets $E_1 \subseteq A \subseteq E_2$ such that $v(E_2 \backslash E_1) < \epsilon$.

Definition 3.6.2 A set $N \subset X = \bar{Q}_N$ is called a *Jordan null set* if and only if $N \in \mathfrak{J}$ and $v(N) = 0$.

Theorem 3.6.2 *A Lebesgue* null *set* $F \subset X = \bar{Q}_N$ *that is* closed *is also a Jordan null set.*

Proof: It suffices to prove that $F \in \mathfrak{J}$. It is easy to see that if $\epsilon > 0$, there exists a set G that is open in $\mathbb{R}^n$ and such that $F \subset G$ and $l(G) < \epsilon$. But the set G is a countable union of *open* rectangular blocks. Since F is compact, the Heine-Borel theorem implies that F can be covered with finitely many of these open rectangular blocks with measure no greater than that of G. Thus there exists an $E \supset F$ such that $E \in \mathcal{E}$, and $v(E) = l(E) < \epsilon$. Thus F is a Jordan null set since $\emptyset \subseteq F \subset E$. ∎

Theorem 3.6.3 *If* $A \subseteq X = \bar{Q}_N$, *then* $A \in \mathfrak{J}$ *if and only if the boundary*

$$\partial A = \bar{A} \backslash A^\circ,$$

the difference between the closure and the interior of A, *is a Jordan null set.*

Proof: Suppose first that $A \in \mathfrak{J}$. Since the boundary ∂A is a closed set, it will suffice to show that it is a Lebesgue null set, since that will imply that it is also a Jordan null set. We know that for each $k \in \mathbb{N}$ there exist sets $E_k \subseteq A \subseteq F_k$ such that both E_k and F_k are in $\mathcal{E}$ and

$$l\left(F_k \backslash E_k\right) = l\left(\bar{F}_k \backslash E_k^\circ\right) < \frac{1}{k}.$$

Let

$$F = \bigcap_{k \in \mathbb{N}} \bar{F}_k \text{ and } E = \bigcup_{k \in \mathbb{N}} E_k^o.$$

It follows that $\partial A \subseteq F \backslash E$, and $l(F \backslash E) = 0$. Thus ∂A is a Lebesgue null set and also a Jordan null set.

Now we suppose that ∂A is a Jordan null set. For each $\epsilon > 0$, there exists $E \in \mathcal{E}$ such that $E^o \supseteq \partial A$, with $l(E) < \epsilon$. We seek to prove that $A \in \mathfrak{J}$. Denote $X = Q^N$ and

$$E^c = X \backslash E^o = \bigcup_{i=1}^{K} B_k \in \mathcal{E}.$$

Suppose that one of the rectangular blocks B_k contained a point $p \in A$ and also a point $q \in A^c$. Then it would follow that the straight line segment from p to q must contain a boundary point of A, which contradicts the hypothesis that $\partial A \subseteq E$. (See Exercise 3.32.) Now let E_1 be the union of the blocks B_k that lie inside A, and let E_2 be the union of the blocks that lie outside A. Then the elementary set $\overline{E_2^c} \backslash E_1^o = E$, and this means that $E_1 \subseteq A \subseteq \overline{E_2^c}$ and $l\left(\overline{E_2^c} \backslash E_1\right) < \epsilon$. Thus $A \in \mathfrak{J}$. ∎

EXERCISES

3.31   Let $A = \mathbb{Q} \cap [0, 1]$, the set of all rational numbers in $[0, l)$. Prove that ∂A is neither a Lebesgue null set nor a Jordan null set, so $A \notin J$. On the other hand, show that $A \in \mathfrak{B}$, the family of Borel sets.

3.32   In $\mathbb{R}^n$, show that if a rectangular block B contains both a point of the set E and a point of the complementary set $E^c = \mathbb{R}^n \backslash E$, then B contains a point in ∂E, the boundary of E.

CHAPTER 4

MEASURABLE FUNCTIONS

If X is a set and $\mathfrak{A} \subseteq \mathfrak{P}(X)$ is a σ-field, then $(X, \mathfrak{A})$ is called a *measurable space*. If μ is a countably additive measure defined on $\mathfrak{A}$, then $(X, \mathfrak{A}, \mu)$ is called a *measure space*. In this chapter we will introduce the family of *measurable functions* for which we will seek to define the Lebesgue integral. We will prove the very important fact that pointwise limits of measurable functions must be measurable. This is encouraging because pointwise limits of Riemann integrable functions need not be Riemann integrable.[37]

4.1 MEASURABLE FUNCTIONS

Definition 4.1.1 Let $(X, \mathfrak{A}, \mu)$ be a measure space.

  i. If $f : X \to \mathbb{R}$, we say that f is $\mathfrak{A}$-*measurable* provided that

$$f^{-1}(-\infty, a) = \{x \in X \mid f(x) < a\} \in \mathfrak{A}$$

---

[37]See Example 1.3.

Measure and Integration: A Concise Introduction to Real Analysis. By Leonard F. Richardson
Copyright © 2009 John Wiley & Sons, Inc.

for all $a \in \mathbb{R}$.[38]

  ii. If $f : X \to \mathbb{C}$, the complex numbers, we write $f(x) = u(x) + iv(x)$ for real-valued functions $u = \Re f$ and $v = \Im f$. We say that f is $\mathfrak{A}$-*measurable* provided that both u and v are $\mathfrak{A}$-measurable.

  iii. If $f : X \to S$, where S is a *topological space*, we say that f is $\mathfrak{A}$-measurable provided that $f^{-1}(G) \in \mathfrak{A}$ for every *open* set $G \subseteq S$.

The reader may note correctly that the concepts of measurability presented above have a formal similarity to the definition of continuity. A function between topological spaces is called *continuous* provided that the inverse image of each open set is open. Of course, for measurability, the inverse image of an open set must be measurable, and measurability is by no means synonymous with being open. In $\mathbb{R}^n$, every open set is Lebesgue measurable, but the converse is clearly false.

The following exercises establish the equivalence of the three different forms of the definition of measurability for functions that are presented in Definition 4.1.1.

EXERCISES

4.1 Let $(X, \mathfrak{A}, \mu)$ be a measure space. Use the steps below to show that the concepts of measurability in Definition 4.1.1 for both real and complex-valued functions are consistent with the concept of measurability for a topological space valued function.

  a) Show that if $f : X \to \mathbb{R}$ is $\mathfrak{A}$-measurable, then $f^{-1}(G) \in \mathfrak{A}$ for every open set $G \subseteq \mathbb{R}$.

  b) Show that $f : X \to \mathbb{R}$ satisfies the definition of measurability if and only if $f^{-1}(G)$ is measurable for each open set $G \subseteq \mathbb{R}$. Do the same for complex-valued functions f.

4.2 Let $(X, \mathfrak{A}, \mu)$ be a measure space. Let $f : X \to \mathbb{R}^n$, the latter space being equipped with its usual Euclidean topology and the Borel field generated by the open sets.

  a) Show that the family

$$\mathfrak{S} = \left\{ A \in \mathfrak{P}(\mathbb{R}^n) \,\middle|\, f^{-1}(A) \in \mathfrak{A} \right\}$$

is a σ-field.

  b) Prove that f is measurable if and only if $f^{-1}(B) \in \mathfrak{A}$ for each Borel set B.[39]

4.1.1 Baire Functions of Measurable Functions

It will be important to know that many combinations of measurable functions and many functions of measurable functions are again measurable. To investigate this, we need the following definition.

---

[38]This definition is motivated by Section 1.3.
[39]For information about f^{-1} of a measurable set, see Exercise 7.14.

Definition 4.1.2 Let $(X, \mathfrak{A}, \mu)$ be any measure space arising from the Hopf Extension Theorem, applied to some measure on a *field* $\mathfrak{E}$. If $f : X \to S$, where S is a *topological space*, we say that f is a *Baire function* provided that

$$f^{-1}(G) \in \mathbb{B}(\mathfrak{E}),$$

the Borel field generated by $\mathfrak{E}$, for each *open* set $G \subseteq S$. Baire functions may also be called *Borel measurable* functions.

Clearly, every Baire function is measurable and every continuous function (from $\mathbb{R}^n$ to S) is a Baire function. The indicator function of a measurable set that is not a Borel set would be an example of a measurable function that is not a Baire function. The indicator function of a nontrivial proper Borel subset of $\mathbb{R}^n$ is an example of a Baire function that is not continuous.

Theorem 4.1.1 *Suppose each of the functions $f_1, f_2, \ldots, f_n$ is an $\mathfrak{A}$-measurable real-valued function defined on X. Let $\Phi : \mathbb{R}^n \to \mathbb{R}$ be a Baire function. Then $F = \Phi(f_1, f_2, \ldots, f_n)$ is an $\mathfrak{A}$-measurable function defined on X.*

Proof: We need to show that for each open set $G \subseteq \mathbb{R}$ we have $F^{-1}(G) \in \mathfrak{A}$. Denote $f = (f_1, f_2, \ldots, f_n) : X \to \mathbb{R}^n$. We claim that f is measurable. In fact, each open set $O \subseteq \mathbb{R}^n$ can be written as a union of countably many open blocks of the form

$$B = \prod_{i=1}^{n} (a_i, b_i),$$

which is a Cartesian product of n open intervals. Thus

$$f^{-1}(B) = \bigcap_{i=1}^{n} f_i^{-1}(a_i, b_i) \in \mathfrak{A}.$$

Hence f is $\mathfrak{A}$-measurable from X to $\mathbb{R}^n$, since $\mathfrak{A}$ is a σ-field and is thus closed under countable unions.

Then

$$F^{-1}(G) = (\Phi \circ f)^{-1}(G) = f^{-1}\left(\Phi^{-1}(G)\right)$$

for each open set $G \subseteq \mathbb{R}$. Since $\Phi^{-1}(G)$ is a Borel set, the theorem is true by the result of Exercise 4.2. ∎

We remark that in Theorem 4.1.1 it would have sufficed to have Φ defined on a set $D \subseteq \mathbb{R}^n$ provided that $(f_1, \ldots, f_n) : X \to D$.

Remark 4.1.1 It follows from Theorem 4.1.1 that such combinations of measurable functions as the following must be measurable:

- $c_1 f_1 + c_2 f_2$, where c_1 and c_2 are constants.

- $f_1 \cdot f_2$, the product of two measurable functions.

- $\frac{f_1}{f_2}$, where f_2 is nowhere zero.

In particular, $-f_1$ is also measurable. Thus f is measurable if and only if

$$\{x \mid f(x) > \alpha\} \in \mathfrak{A}$$

for every $\alpha \in \mathbb{R}$. Of course, this can be seen also from the fact that f is measurable if and only if f^{-1} maps open sets to measurable sets.

Corollary 4.1.1 *If f_1 and f_2 are $\mathfrak{A}$-measurable real-valued functions, then each of the following functions is $\mathfrak{A}$-measurable:*

    i. $\max(f_1, f_2) = f_1 \vee f_2$, *where* $f_1 \vee f_2(x) = \max(f_1(x), f_2(x))$ *for each* $x \in X$.

    ii. $\min(f_1, f_2) = f_1 \wedge f_2$, *where* $f_1 \wedge f_2(x) = \min(f_1(x), f_2(x))$ *for each* $x \in X$.

    iii. $f^+ = f \vee 0$, *known as the* positive part *of f.*

    iv. $f^- = (-f) \vee 0$, *known as the* negative part *of f.* *(The reader should note that the* negative part *of a real-valued function is* positive.)

    v. $|f| = f^+ + f^-$.

Proof: It suffices to observe that $\max(x_1, x_2)$ and $\min(x_1, x_2)$ are both continuous functions from $\mathbb{R}^2$ to $\mathbb{R}$, making each of these a Baire function. For the last part, we use the fact that $\Phi(x_1, x_2) = x_1 + x_2$ is continuous and thus a Baire function.   ∎

EXERCISE

4.3   Show that it is possible for $f : \mathbb{R} \to \mathbb{R}$ to be nonmeasurable and still have $|f|$ measurable.

4.2   LIMITS OF MEASURABLE FUNCTIONS

In the study of pointwise limits of measurable functions and integrable functions, we will consider sequences of functions for which $f_n(x)$ diverges to $\pm\infty$ for some values of x. Thus it is helpful to extend the concept of real numbers to the set $\mathbb{R}^* = \mathbb{R} \cup \{\pm\infty\}$ of *extended real numbers*. Measurability for an extended real-valued function means that for each $\alpha \in \mathbb{R}$, the set

$$f^{-1}[\alpha, \infty] = \{x \mid f(x) \geqslant \alpha\} \in \mathfrak{A},$$

and conversely.

Theorem 4.2.1 *Let $(X, \mathfrak{A}, \mu)$ be a measure space and let $\{f_n \mid n \in \mathbb{N}\}$ be any sequence of measurable functions from X to $\mathbb{R}^*$. Then each of the five functions defined as follows is $\mathfrak{A}$-measurable:*

i. $f_*(x) = \inf\{f_n(x) \mid n \in \mathbb{N}\}$ *for all* $x \in X$.

ii. $f^*(x) = \sup\{f_n(x) \mid n \in \mathbb{N}\}$ *for all* $x \in X$.

iii. $\underline{f}(x) = \liminf f_n(x)$ *for all* $x \in X$.

iv. $\overline{f}(x) = \limsup f_n(x)$ *for all* $x \in X$.

v. $f(x) = \lim_{n \to \infty} f_n(x)$ *provided the limit exists for all* $x \in X$.

Proof: For the first part, we observe that each $a \in \mathbb{R}$,

$$f_*^{-1}[\alpha, \infty] = \left\{x \,\middle|\, \inf_n f_n(x) \geq \alpha\right\}$$
$$= \bigcap_{n \in \mathbb{N}} f_n^{-1}[\alpha, \infty] \in \mathfrak{A}.$$

Thus f_* is $\mathfrak{A}$-measurable. Since

$$f^*(x) \equiv -\inf_n(-f_n(x)),$$

it follows that f^* is $\mathfrak{A}$-measurable as well. Note next that since

$$i_n = \inf\{f_k(x) \mid k \geq n\}$$

is an increasing sequence of extended real numbers i_n, we have

$$\liminf f_n(x) = \underline{f}(x) = \sup\left\{\inf\{f_k(x) \mid k \geq n\} \,\middle|\, n \in \mathbb{N}\right\},$$

so that $\underline{f}$ is $\mathfrak{A}$-measurable, being the supremum of a sequence of measurable functions given as infima. Also,

$$\limsup f_n(x) = \overline{f}(x) = \inf\left\{\sup\{f_k(x) \mid k \geq n\} \,\middle|\, n \in \mathbb{N}\right\},$$

with the result that $\overline{f}$ is $\mathfrak{A}$-measurable. Finally, we note that

$$f(x) = \lim_{n \to \infty} f_n(x)$$

exists if and only if $\overline{f}(x) = \underline{f}(x)$. Thus f is $\mathfrak{A}$-measurable provided that the pointwise limit exists on X. ∎

EXERCISE

4.4 Suppose $f_n : X \to \mathbb{R}$ is a measurable function for each $n \in \mathbb{N}$, where $(X, \mathfrak{A}, \mu)$ is a measure space. Prove that the set

$$S = \left\{x \,\middle|\, \lim_{n \to \infty} f_n(x) \text{ exists}\right\}$$

is a measurable set.

The reader should be able to give examples of pointwise convergent sequences of continuous functions for which the limit is not continuous and examples of pointwise convergent sequences of Riemann integrable functions for which the limit is not Riemann integrable. We see already that measurability must be a valuable concept for pointwise convergence since pointwise convergence does preserve measurability.

Definition 4.2.1 Let $(X, \mathfrak{A}, \mu)$ be a measure space, and let $\mathfrak{N}$ be the set of all *null sets*. Suppose $P(x)$ is a proposition for each $x \in X$. Then we say that $P(x)$ is valid

$$\mathfrak{A}\text{-}\mu \quad \text{almost everywhere}$$

if and only if there is a set $N \in \mathfrak{N}$ such that x has the property P for all $x \in X \backslash N$. This is commonly expressed as μ-*almost everywhere*, or as *almost everywhere*, provided that there will be no confusion as to which measure μ or σ-algebra $\mathfrak{A}$ is in use.

Remark 4.2.1 Given a measurable function f, it is common to see a set-theoretic notation such as this: Let

$$B = \inf\{K \in \mathbb{R} \mid |f(x)| \leqslant K \text{ a.e.}\}.$$

In this notation, each of the numbers K is called an *essential upper bound* of f, meaning that $|f(x)|$ is bounded above by K *almost everywhere*. If $B < \infty$, then we denote

$$B = \|f\|_x,$$

the *essential sup-norm* of f.

Corollary 4.2.1 *Let f_n be a sequence of measurable functions on a complete measure space $(X, \mathfrak{A}, \mu)$. Suppose $f_n(x) \to f(x)$ almost everywhere on X. Then the function f is measurable.*

Proof: This follows from Theorem 4.2.1. It is understood in this context that although the function f is defined by the given limit, except on a null set N, f may have arbitrary values on N itself. Then the completeness of the measure space tells us that the resulting function f remains $\mathfrak{A}\text{-}\mu$ measurable regardless of how values are assigned to f within the null set N.   ∎

EXERCISES

4.5   Suppose $f : X \to \mathbb{R}^*$ is a measurable function that has finite values almost everywhere on the measure space $(X, \mathfrak{A}, \mu)$, where $\mu(X) > 0$. Prove that there is a set of positive measure on which f is bounded.

4.6   Suppose the measurable function $f : \mathbb{R}^n \to \mathbb{R}$ has the special property that for each fixed vector $c \in \mathbb{R}^n$, the translation of f given by $f_c(x) = f(x + c)$ is equal almost everywhere to $f(x)$ itself. That is, for each value of $c \in \mathbb{R}^n$, we have

$$f_c(x) = f(x)$$

for almost all x. Prove that $f(x)$ is equal almost everywhere to a constant function. (Hints: Consider both the sum and the terms of

$$\sum_{n \in \mathbb{Z}} l(f^{-1}[n, n+1)).$$

Apply Exercise 3.26 to select the special value of n. *Divide and conquer!*)

4.3 SIMPLE FUNCTIONS AND EGOROFF'S THEOREM

In this section we will consider both *uniform convergence* and what we will call *almost uniform convergence* of sequences of measurable functions. We begin with a useful definition that generalizes the notion of a step function from advanced calculus.

Definition 4.3.1 A function $f : X \to \mathbb{R}$ is called $\mathfrak{A}$-*simple* (or *simple* if there will be no confusion regarding the σ-field $\mathfrak{A}$ that is under consideration) if and only if f is $\mathfrak{A}$-measurable *and*

$$\{f(x) \mid x \in X\}$$

is a finite set. The class of simple functions is denoted by $\mathfrak{S}$.

Thus, f is simple if and only if we have finitely many sets

$$A_i = f^{-1}(\alpha_i) \in \mathfrak{A}, \ i = 1, \dots, n$$

such that

$$X = \bigcup_{i=1}^{n} A_i \ \text{ and } \ f(x) = \sum_{i=1}^{n} \alpha_i 1_{A_i}(x),$$

where 1_A denotes the *indicator function* of A. That is,

$$1_A(x) = \begin{cases} 1 \text{ if } x \in A, \\ 0 \text{ if } x \notin A. \end{cases}$$

Theorem 4.3.1 *Every* bounded, $\mathfrak{A}$-*measurable, real-valued function f is the* uni*form limit of a monotone nondecreasing sequence of simple functions and also of a monotone nonincreasing sequence of simple functions.*

Proof: Define

$$f_n(x) = \frac{k}{2^n} \text{ if } x \in A_k = f^{-1}\left[\frac{k}{2^n}, \frac{k+1}{2^n}\right).$$

Since f is bounded, there are only finitely many values of k for which $A_k \neq \emptyset$. Thus f_n achieves only finitely many values and is $\mathfrak{A}$-simple. Moreover,

$$\|f_n - f\|_{\sup} \leq \frac{1}{2^n} \to 0$$

so that $f_n \to f$ uniformly on X.

It is left to the reader to verify that f_n is in fact a *monotone nondecreasing* sequence of simple functions. Monotonicity follows from the method of *interval bisection* that is used in the definition of the sequence f_n. The reader should also produce and verify the properties of a similar nonincreasing sequence of simple functions. This can be done by applying the preceding method to the function $-f$.   ∎

It is worth noting that even the pointwise limit of a sequence of simple functions must be measurable. Thus a bounded function is measurable if and only if it is the limit of a monotone sequence of simple functions.

EXERCISES

4.7 Prove that every $\mathfrak{A}$-measurable function $f : X \to \mathbb{R}$ is the pointwise limit of a sequence of $\mathfrak{A}$-simple functions. Give an example to show that uniformity of convergence cannot be required in this exercise. (Hint: Proceed as in the proof of Theorem 4.3.1, but for each n define f_n using $|k| \leqslant n2^n$.)

4.8 Let $f : \mathbb{R}^n \to \mathbb{R}$ be Lebesgue measurable. Prove that f is equal almost everywhere to a *Borel* measurable function. (Hint: The function f is the pointwise limit of a sequence $\phi_n \in \mathfrak{S}$, the vector space of simple functions. Modify the functions ϕ_n suitably.)

Exercise 4.7 shows that if the measurable function f is not bounded, then we cannot expect uniform approximation by simple functions. The following theorem provides interesting additional insight into what can be guaranteed.

Theorem 4.3.2 (Egoroff) *Let* $(X, \mathfrak{A}, \mu)$ *be a measure space of* finite *total measure. Let* f_n *be* $\mathfrak{A}$-*measurable for each* $n \in \mathbb{N}$ *and suppose that* $f_n \to f$ *pointwise on* X. *Then, for each* $\eta > 0$, *there exists a set* $B \in \mathfrak{A}$ *such that* $\mu(B) < \eta$ *and such that* $f_n \to f$ *uniformly on* $X \backslash B$.

Proof: Note that f must be measurable because of the hypotheses. Let $\epsilon > 0$ and define

$$A_n(\epsilon) = \left\{ x \in X \, \middle| \, \exists p \geqslant n \text{ such that } |f_p(x) - f(x)| \geqslant \epsilon \right\}$$
$$= \bigcup_{p \geqslant n} \left\{ x \in X \, \middle| \, |f_p(x) - f(x)| \geqslant \epsilon \right\}.$$

This set is measurable since each f_p is measurable. Thus $A_n \in \mathfrak{A}$ for each $n \in \mathbb{N}$. Moreover,

$$A_1(\epsilon) \supseteq A_2(\epsilon) \supseteq \ldots \supseteq A_n(\epsilon) \supseteq \ldots$$

is a decreasing nest with *empty intersection* because f_n converges pointwise on X. By Exercise 2.19,

$$\mu\big(A_n(\epsilon)\big) \to 0$$

as $n \to \infty$, since $\mu(X) < \infty$.

For each $k \in \mathbb{N}$ there exists $n_k \in \mathbb{N}$ such that

$$\mu\left(A_{n_k}\left(\frac{1}{k}\right)\right) < \frac{\eta}{2^k}$$

and such that $n_1 < n_2 < \dots$. Let

$$B = \bigcup_{k=1}^{\infty} A_{n_k}\left(\frac{1}{k}\right),$$

so that $\mu(B) < \eta$. If $\epsilon > 0$, we can pick k such that $\frac{1}{k} < \epsilon$. If $x \notin B$ and $n \geqslant n_k$, then $|f_n(x) - f(x)| < \epsilon$. Thus $f_n \to f$ uniformly on $X \backslash B$. ∎

We remark that Egoroff's theorem is often paraphrased informally as follows: *Pointwise convergence is almost uniform convergence*. It is necessary, however, to understand the precise meaning of that expression as being the statement of Egoroff's theorem. It is especially important to understand that the set B must normally have strictly positive measure, albeit small measure. We give an elementary example to explain this.

■ **EXAMPLE 4.1**

Let

$$f_n(x) = \left(1 - x^2\right)^n$$

on $[-1, 1]$ for each $n \in \mathbb{N}$, using the underlying Lebesgue measure space $([-1, 1], \mathcal{L}, l)$. Then $f_n \to 1_{\{0\}}$ pointwise, but not uniformly, on $[-1, 1]$. If we let $B = (-r, r)$ for a small positive number r, then it is easy to see that $f_n \to 0$ uniformly on $[-1, 1] \backslash B$. In fact,

$$\|f_n - 0\|_{\sup} = \left(1 - r^2\right)^n \to 0,$$

with the sup-norm being calculated on $[-1, 1] \backslash B$.

For a set B to have the effect calculated above, it must include some interval around 0 so as to exclude *all* sequences different from 0 but converging to 0. Thus B cannot be replaced by a null set.

EXERCISE

4.9 Give an example on the real line to show that pointwise convergence of a sequence of measurable functions need not imply uniform convergence off some set of small measure. Thus finiteness of the measure space is necessary in Egoroff's theorem.

4.3.1 Double Sequences

Theorem 4.3.3 *Let $(X, \mathfrak{A}, \mu)$ be a finite measure space. Let $\{f_{i,j} \mid i, j = 1, 2, \dots\}$ be a double sequence of $\mathfrak{A}$-measurable functions defined on X such that*

$$\lim_{j \to \infty} f_{i,j}(x) = f_i(x), \ \forall x \in X$$

and
$$\lim_{i \to \infty} f_i(x) = f(x), \ \forall x \in X.$$
Then there exists an increasing sequence n_i of positive integers such that
$$\lim_{i \to \infty} f_{i,n_i}(x) = f(x)$$
$\mathfrak{A}$-μ *almost everywhere on X.*

Proof: We can diagram the hypotheses conveniently in the form
$$f_{i,j}(x) \stackrel{(j \to \infty)}{\longrightarrow} f_i(x)$$
$$\Big\downarrow (i \to \infty)$$
$$f(x).$$
There exist $B_1 \in \mathfrak{A}$ and $n_1 \in \mathbb{N}$ such that $\mu(B_1) < \frac{1}{2}$ and
$$|f_{1,n_1}(x) - f_1(x)| < \frac{1}{2}$$
for all $x \in X \backslash B_1$ because of Egoroff's theorem. Similarly, there exist $B_i \in \mathfrak{A}$ and $n_i \in \mathbb{N}$ such that $\mu(B_i) < \dfrac{1}{2^i}$ and
$$|f_{i,n_i}(x) - f_i(x)| < \frac{1}{2^i}$$
for all $x \in X \backslash B_i$. Now let
$$B = \bigcap_{n=1}^{\infty} \bigcup_{i=n}^{\infty} B_i.$$
Observe that $\mu(B) = 0$. Also, B is the set of all those x that lie in infinitely many of the sets B_i. Hence if $x \in X \backslash B$, it follows that x lies in at most finitely many of the sets B_i, for which reason $|f_{i,n_i}(x) - f(x)| \to 0$ as $i \to \infty$. ∎

Remark 4.3.1 We defined the concept of Baire function in Definition 4.1.2. There is an alternative, equivalent definition that could be given for functions defined on the real line. The Baire class B_0 is the class of continuous functions. B_1 is the class of all pointwise limits of functions in B_0. For each $\alpha < \Omega$, the smallest uncountable ordinal number, we can define B_α to be the set of all pointwise limits of functions belonging to lower Baire classes. The family of all Baire functions as defined in Definition 4.1.2 can be shown to be
$$\bigcup_{\alpha < \Omega} B_\alpha.$$
The detailed explanation can be found in the book by Casper Goffman [8]. One part of the significance of the preceding theorem is that there are functions of Baire class B_2 that are not of class B_1. Thus almost-everywhere convergence is the best that can be expected in the theorem.

4.3.2   Convergence in Measure

There is a concept called *convergence in measure* of a sequence of measurable functions f_n to f that is especially useful in the theory of probability. In that context, it is helpful to to know that the probability of a random variable f_n differing from the random variable f by more than ϵ is very small.

Definition 4.3.2   On a measure space $(X, \mathfrak{A}, \mu)$, a sequence of measurable functions f_n is said to *converge in measure* to a measurable function f provided that for each $\epsilon > 0$ there exists $N \in \mathbb{N}$ such that $n \geqslant N$ implies that

$$\mu \left\{ x \,\middle|\, |f_n(x) - f(x)| \geqslant \epsilon \right\} < \epsilon.$$

It follows readily from the definition that $f_n \to f$ in measure if and only if for each $\epsilon > 0$ and each $\eta > 0$ there exists $N \in \mathbb{N}$ such that $n \geqslant N$ implies

$$\mu \left\{ x \,\middle|\, |f_n(x) - f(x)| \geqslant \eta \right\} < \epsilon.$$

Thus the definition is phrased as it is for simplicity. One gains nothing that is not already there if one uses two criteria, ϵ and η.

EXERCISES

4.10   Let $(X, \mathfrak{A}, \mu)$ be a measure space for which $\mu(X) < \infty$. Suppose f_n is a sequence of measurable functions such that $f_n \to f$ almost everywhere. Prove that $f_n \to f$ in measure. (Hints: Theorem 4.3.2 is helpful. You may assume *either* that f is measurable *or* that the measure space is complete, explaining why either assumption has the same effect.)

4.11   Give an example of a sequence of Lebesgue measurable functions $f_n \to 0$ in measure on $[0, 1]$ for which the sequence of numbers $f_n(x)$ fails to converge to zero for any $x \in [0, 1]$.

Exercise 4.11 adds to the significance of the theorem below. The exercise explains why, in the following theorem, we will need need to pass to a subsequence that is sufficiently *rapidly* convergent in measure to guarantee pointwise convergence almost everywhere.

Theorem 4.3.4 *Let f and f_n be measurable functions on a finite measure space $(X, \mathfrak{A}, \mu)$ for all n. Suppose that $f_n \to f$ in measure. Then there exists a subsequence f_{n_ν} that converges to f almost everywhere as $\nu \to \infty$.*

Proof: By hypothesis, for each $\nu \in \mathbb{N}$ there exists $n_\nu \in \mathbb{N}$ such that $n \geqslant n_\nu$ implies that

$$\mu \left\{ x \,\middle|\, |f_n(x) - f(x)| \geqslant \frac{1}{2^\nu} \right\} < \frac{1}{2^\nu}.$$

The difficulty in establishing convergence pointwise almost everywhere is that these sets can slide around and cover a big region as we vary $n \geqslant n_\nu$. Thus we define the set

$$E_\nu = \left\{ x \middle| |f_{n_\nu}(x) - f(x)| \geqslant \frac{1}{2^\nu} \right\}$$

for the individual function f_{n_ν}, and we do this for each value of ν. Define

$$S = \limsup E_\nu = \bigcap_{k=1}^{\infty} \bigcup_{\nu=k}^{\infty} E_\nu.$$

It is easy to check that S is a null set. Moreover, $x \notin S$ if and only if x lies in only finitely many of the sets E_ν. Thus if $x \notin S$, we know that for sufficiently big values of ν corresponding to x, we have $x \notin E_\nu$. This implies that

$$|f_{n_\nu}(x) - f(x)| < \frac{1}{2^\nu}.$$

It follows that for $x \notin S$ the sequence $f_{n_\nu}(x) \to f(x)$ as $\nu \to \infty$. ∎

4.4 LUSIN'S THEOREM

Lusin's theorem is paraphrased often as stating that a measurable function on $\mathbb{R}^p$ is *almost a continuous function*. Such phrasings can be useful as reminders of theorems that could help us in certain situations. But it is very important to remember to interpret the paraphrasing of Lusin's theorem as meaning exactly what the theorem itself states.

Theorem 4.4.1 (Lusin) *Let $f : X \to \mathbb{R}$ be a measurable function defined on a Lebesgue measurable set $X \subset \mathbb{R}^p$, for which the Lebesgue measure $l(X)$ is finite. Then for each $\eta > 0$ there exists a* compact *subset $K \subseteq X$ such that $l(X \backslash K) < \eta$ and such that $f\big|_K$, the* restriction of f to K, is continuous *on K.*

Proof: We will undertake four restrictions of domain in order to reach a compact set K on which $f\big|_K$ is continuous.

i. We wish to restrict f to a bounded subset of X so that the closed approximations of measurable sets from within will be compact. We can do this as follows. Since

$$l(X) = \lim_{k \to \infty} l\left(X \cap [-k, k]^p \right),$$

there exists a closed cube $Q = [-K, K]^p \subset \mathbb{R}^p$ large enough so that

$$l(X) \geqslant l(Q \cap X) > l(X) - \frac{\eta}{8}.$$

ii. We know from Exercise 4.7 that f is the pointwise limit of functions $f_n \in \mathfrak{S}$, the class of simple Lebesgue measurable functions. Write

$$f_n = \sum_{i=1}^{p_n} \alpha_i^n 1_{A_i^n},$$

a linear combination of indicator functions of *disjoint* measurable sets A_i^n, with $X = \bigcup_{i=1}^{p_n} A_i^n$. (The superscripts are labels only—not exponents.) By Exercise 3.3, for each i and n there exists a compact set $K_i^n \subseteq Q \cap A_i^n$ such that

$$l((Q \cap A_i^n) \backslash K_i^n) < \frac{\eta}{p_n 2^{n+1}}.$$

Each function f_n is continuous on K_i^n because it is constant there. Note also that the cluster points of K_i^n and those of K_j^n for $j \neq i$ must be distinct, since both sets are compact subsets of their respective disjoint measurable sets. That is, each set contains all its limit points and the two compact sets are disjoint. Thus f_n is continuous also on

$$K^n = \bigcup_{i=1}^{p_n} K_i^n.$$

Moreover

$$l(X \backslash K^n) < \frac{\eta}{8} + \sum_{i=1}^{p_n} \frac{\eta}{p_n 2^{n+1}} = \frac{\eta}{8} + \frac{\eta}{2^{n+1}}.$$

iii. Define another compact set

$$K^* = \bigcap_{n=1}^{\infty} K^n$$

so that

$$l(X \backslash K^*) < \frac{\eta}{8} + \frac{\eta}{2} = \frac{5\eta}{8}.$$

The functions f_n are continuous on K^* and the sequence $f_n \to f$ pointwise on K^*.

iv. By Egoroff's theorem there exists a measurable set $B \subseteq K^*$ with $l(B) < \frac{\eta}{4}$ and $f_n \to f$ *uniformly* on $K^* \backslash B$. Since the set $K^* \backslash B$ is measurable, there exists a compact set $K \subseteq K^* \backslash B$ such that

$$l((K^* \backslash B) \backslash K) < \frac{\eta}{8}$$

which implies that $l(X \backslash K) < \eta$ and f is continuous on K. ∎

Lusin's theorem tells us that the *restriction* $f|_K$ is a continuous function. That is, $f \in C(K, \mathbb{R})$. In other words, f is continuous as a function defined only on the restricted domain, K. But f need not be continuous at any $k \in K$ as a function defined on X. This is relevant to Exercise 4.12.

EXERCISES

4.12   Let f be the indicator function of the set of all irrational numbers in the interval $X = [0, 1]$.

   a) Show that f is nowhere continuous on $[0, 1]$.

   b) Let $\eta > 0$ and find a set B of measure less than η such that f is continuous on $K = X \backslash B$ and such that K is compact.

4.13   Let $f : [a, b] \to \mathbb{R}$ be a measurable function. Let $\eta > 0$ and $\epsilon > 0$. Prove that there exists a measurable set B such that $l(B) < \eta$ and a polynomial p such that

$$\sup \{ |f(x) - p(x)| \, | \, x \in [a, b] \backslash B \} < \epsilon.$$

(Hint: Apply the Tietze Extension theorem and the Weierstrass Approximation theorem.)

4.14   Use Lusin's theorem to prove that a measurable homomorphism f of the additive group of real numbers into itself must be continuous.[40] That is, prove that if f is measurable and if

$$f(x + y) \equiv f(x) + f(y),$$

then f must be continuous. (Hints: Consider a compact subset $K \subset [0, 1]$ that *nearly fills* $[0, 1]$ in the sense of Lebesgue measure and such that f is uniformly continuous on K. Prove that if h is sufficiently small, then $K \cap (K + h)$ is nonempty. Prove that f is continuous at 0.[41])

---

[40]This problem appears again in the present book as Exercise 7.7, on page 131. See how the problem is expressed in that exercise, and read the historical footnote following it. For that second introduction of the problem, an integration-based solution is suggested.

[41]This method was presented in a paper by S. Banach [1].

CHAPTER 5

THE INTEGRAL

We will introduce the concept of the integral, beginning with the integration of *special simple functions* and some easy theorems.

5.1 SPECIAL SIMPLE FUNCTIONS

Throughout this section we let $(X, \mathfrak{A}, \mu)$ be a measure space. We introduce below the concept of a *special simple function* which generalizes the concept of a step function in the study of the Riemann integral. Recall that a simple function is a measurable function that achieves only finitely many distinct values, all of them real.

Definition 5.1.1 The set $\mathfrak{S}_0$ of *special simple functions* consists of those real-valued simple functions $f \in \mathfrak{S}$ denoted by

$$f = \sum_{i=1}^{n} \alpha_i 1_{A_i},$$

for which $\alpha_i \neq 0$ implies that $\mu(A_i) < \infty$.

Measure and Integration: A Concise Introduction to Real Analysis. By Leonard F. Richardson
Copyright © 2009 John Wiley & Sons, Inc.

Observe that for a special simple function, if $\mu(A_i) = \infty$, then we must have $\alpha_i = 0$. Next, we define the integral of a special simple function in the most natural way. The definition will clarify why we need to assume that $\mu(A_i) < \infty$ if $\alpha_i \neq 0$.

Definition 5.1.2 If $f = \sum_{i=1}^{n} \alpha_i 1_{A_i} \in \mathfrak{S}_0$, we define

$$\int_X f \, d\mu = \sum_{i=1}^{n} \alpha_i \mu(A_i). \tag{5.1}$$

We adopt the convention, in Equation (5.1), that if $\mu(A_i) = \infty$, so that $\alpha_i = 0$, then the corresponding summand, $\alpha_i \mu(A_i)$, is to be counted as *zero* in the sum.

We leave it for the reader to check that the integral is well defined on $\mathfrak{S}_0$, despite the fact that the decomposition of such a function as a linear combination of indicator functions of measurable sets is not unique.

Theorem 5.1.1 (Linearity of the Integral) *The space* $\mathfrak{S}$ *is a vector space and the space* $\mathfrak{S}_0$ *is a vector space. Let* $I : \mathfrak{S}_0 \to \mathbb{R}$ *be defined by*

$$I(f) = \int_X f \, d\mu.$$

Then I is a linear functional, *meaning that* $I(\alpha f + g) = \alpha I(f) + I(g)$ *for all f and g in* $\mathfrak{S}_0$ *and* $\alpha \in \mathbb{R}$.

Proof: To prove that $\mathfrak{S}_0$ and $\mathfrak{S}$ are vector spaces, it is necessary only to prove closure under subtraction and scalar multiplication. If f and g are in either of these two spaces, we can write

$$f = \sum_{i=1}^{m} \alpha_i 1_{A_i} \text{ and } g = \sum_{j=1}^{n} \beta_j 1_{B_j}.$$

We will require for convenience in this proof that

$$\bigcup_{i=1}^{m} A_i = X = \bigcup_{j=1}^{n} B_j.$$

Hence we can write

$$\alpha f + g = \sum_{\substack{1 \leq i \leq m \\ 1 \leq j \leq n}} (\alpha \alpha_i + \beta_j) 1_{A_i \cap B_j}.$$

It follows by direct application of the definition of I that

$$I(\alpha f + g) = \alpha I(f) + I(g).$$

∎

The following exercises and theorems present basic properties of the integral of a special simple function. The proofs will be quite simple.

EXERCISES

5.1   (Monotonicity of the Integral) Let f and g be in $\mathfrak{S}_0$. Prove the following statements.

   a) If $f(x) \geqslant 0$ for all $x \in X$, then $\displaystyle\int_X f \, d\mu \geqslant 0$. *(Positivity of the Integral)*

   b) If $f(x) \leqslant g(x)$ for all $x \in X$, then $\displaystyle\int_X f \, d\mu \leqslant \int_X g \, d\mu$. *(Monotonicity of the Integral)* (Hint: Apply Theorem 5.1.1.)

5.2   If $f \in \mathfrak{S}_0$, then $|f| \in \mathfrak{S}_0$ and $\left| \displaystyle\int_X f d\mu \right| \leqslant \int_X |f| \, d\mu$.

Theorem 5.1.2 (Triangle Inequality) *Let f, g, and h be in $\mathfrak{S}_0$. Define*

$$d(f, g) = \int_X |f - g| \, d\mu.$$

Then $d(f, h) \leqslant d(f, g) + d(f, h)$.

Proof: For each $x \in X$,

$$|f(x) - h(x)| \leqslant |f(x) - g(x)| + |g(x) - h(x)|$$

by the triangle inequality for real numbers. The rest follows from Exercise 5.1.   ∎

Remark 5.1.1 In the vector space $\mathfrak{S}_0$, define $f \sim g$ if and only if $d(f, g) = 0$. We can verify easily that this is an equivalence relation and that this equivalence relation partitions $\mathfrak{S}_0$ into equivalence classes. For each special simple function f, we denote by $\bar{f}$ the equivalence class of f. With the function d serving as a metric, the quotient space $\mathfrak{S}_0/\sim$ of equivalence classes formed from $\mathfrak{S}_0$ is a metric space. Note that the verification that d is a full-fledged metric on the space of equivalence classes includes the fact that if $f(x) \geqslant 0$ for all $x \in X$ and if $\int_X f \, d\mu = 0$, then $f(x) = 0$ almost everywhere.

Definition 5.1.3 Let $f = \sum_{i=1}^n \alpha_i 1_{A_i} \in \mathfrak{S}_0$ and let $A \in \mathfrak{A}$, the σ-field of measurable sets. Then the pointwise product $f 1_A \in \mathfrak{S}_0$, and we define

$$\int_A f \, d\mu = \int_X f 1_A \, d\mu = \sum_{i=1}^n \alpha_i \mu(A \cap A_i).$$

EXERCISES

5.3   (Countable Additivity as a Set Function) If $f \in \mathfrak{S}_0$ and if we have a sequence of *disjoint* measurable sets $E_i \in \mathfrak{A}$, then use the countable additivity of μ to prove that

$$\int_{\dot{\bigcup}_{i \in \mathbb{N}} E_i} f d\mu = \sum_{i \in \mathbb{N}} \int_{E_i} f d\mu.$$

5.4   Let $X = [a, b]$, a closed finite interval, and let $S = \mathbb{Q} \cap X$. Prove that $1_S \in \mathfrak{S}_0$ with respect to Lebesgue measure on X. Find $\int_{[a,b]} 1_S \, dl$ and $\int_{[a,b]} 1_{S^c} \, dl$. Compare with Exercise 1.3.

5.2   EXTENDING THE DOMAIN OF THE INTEGRAL

We are ready to expand the domain of functions that we can integrate. We will begin with the bounded measurable functions having a carrier of finite measure.

Definition 5.2.1 Let $\mathfrak{S}_1 \supset \mathfrak{S}_0$ be the family of functions f that are not necessarily simple but that have the following properties:

  i. The function f is $\mathfrak{A}$-measurable.

  ii. There is a set $A \in \mathfrak{A}$ with $\mu(A) < \infty$ such that $f(x) = 0$ for all $x \notin A$. (The set A is called a *carrier* of f. Property (ii) says that f has a carrier A of finite measure.)

  iii. There is a nonnegative constant $M \in \mathbb{R}$ such that $|f(x)| \leqslant M$ for all $x \in A$.

The definition may summarized in words by saying that $f \in \mathfrak{S}_1$ if and only if f is *bounded and measurable with finite carrier*.

By Theorem 4.3.1, we know that there exist both a monotone nondecreasing sequence $g_n \in \mathfrak{S}_0$ and a monotone nonincreasing sequence $h_n \in \mathfrak{S}_0$ of simple functions, both of which converge *uniformly* on A to f. That is, $\|h_n - f\|_{\sup} \to 0$ and $\|g_n - f\|_{\sup} \to 0$, where

$$\|\phi\|_{\sup} = \sup \{|\phi(x)| \mid x \in A\}$$

is the *sup-norm* of any function ϕ defined on A. From the triangle inequality for the sup-norm, it follows that $\|h_n - g_n\|_{\sup} \to 0$ as $n \to \infty$ as well. Define

$$\Phi(\phi) = \int_A \phi \, d\mu,$$

for each $\phi \in \mathfrak{S}_0$. We know from the monotonicity of the integral of a special simple function that $\Phi(g_n) \leqslant \Phi(h_n)$ for all n, that $\Phi(g_n)$ is a nondecreasing sequence, and that $\Phi(h_n)$ is a nonincreasing sequence. Moreover,

$$\Phi(h_n) - \Phi(g_n) = \Phi(h_n - g_n)$$
$$= \int_A h_n - g_n \, d\mu$$
$$\leqslant \int_A |h_n - g_n| \, d\mu$$
$$\leqslant \int_A \|h_n - g_n\|_{\sup} \, d\mu$$
$$= \mu(A)\|h_n - g_n\|_{\sup} \to 0$$

as $n \to \infty$. Recall that every bounded monotone sequence of real numbers must converge. This argument establishes also that for each monotone nondecreasing sequence $g_n \in \mathfrak{S}_0$ converging uniformly to f and for each monotone nonincreasing sequence $h_n \in \mathfrak{S}_0$ converging uniformly to f, $\lim \Phi(g_n)$ exists and equals $\lim \Phi(h_n)$.

This justifies the following definition.

Definition 5.2.2 If $f \in \mathfrak{S}_1$, we define

$$\int_X f\, d\mu = \lim_{n \to \infty} \Phi(\phi_n),$$

where ϕ_n is any monotone sequence of special simple functions converging uniformly to f on its carrier set A of finite measure.

If $E \in \mathfrak{A}$ and f is as in Definition 5.2.2, then $f \, 1_E \in \mathfrak{S}_1$, and we observe that

$$\int_E f\, d\mu = \int_A f \, 1_E \, d\mu.$$

Corollary 5.2.1 *Let* $f \in \mathfrak{S}_1$ *and let* $s_n \in \mathfrak{S}_0$ *be any* sequence, not *necessarily* monotone, *converging uniformly to* f *on its carrier set* A *of finite measure. Then*

$$\int_X s_n \, d\mu \to \int_X f \, d\mu$$

as $n \to \infty$.

Proof: Let ϕ_n be any *monotonic* sequence, as in Definition 5.2.2. We know that $\int_A \phi_n \, d\mu \to \int_A f \, d\mu$. But

$$\left| \int_X s_n \, d\mu - \int_X \phi_n \, d\mu \right| \leq \int_X \|s_n - \phi_n\|_{\sup} \, d\mu$$

$$\leq \mu(A) \|s_n - \phi_n\|_{\sup} \to 0$$

as $n \to \infty$. This proves that $\lim_{n \to \infty} \int_X s_n \, d\mu$ exists and equals $\int_X f \, d\mu$. ∎

Remark 5.2.1 If $f \in \mathfrak{S}_1$, then

$$\int_X f \, d\mu = \sup \left\{ \int_X \phi \, d\mu \,\middle|\, \phi \in \mathfrak{S}_0, \ \phi \leq f \right\}.$$

The reader can prove this as follows by proving two inequalities. One of the inequalities, that

$$\int_X f \, d\mu \leq \sup \left\{ \int_X \phi \, d\mu \,\middle|\, \phi \in \mathfrak{S}_0, \ \phi \leq f \right\},$$

is immediate. For the other direction, consider any sequence of increasing special simple functions ϕ_n converging uniformly to f. Then the sequence of maximum functions $\phi_n \vee \phi \to f$ in the same way. Now the other inequality follows directly.

EXERCISE

5.5 Prove that the mapping $\Phi : \mathfrak{S}_1 \to \mathbb{R}$ defined by $\Phi(f) = \int_X f \, d\mu$ has the following properties:

    a) $\mathfrak{S}_1$ is a vector space and the integral is *linear*.

    b) The integral is *positive and monotone*, meaning that if $f \in \mathfrak{S}_1$ is everywhere nonnegative, then $\Phi(f) \geqslant 0$. If $f \leqslant g$, then $\displaystyle\int_X f \, d\mu \leqslant \int_X g \, d\mu$.

    c) If $f \in \mathfrak{S}_1$, then $|f| \in \mathfrak{S}_1$ and $\displaystyle\left| \int_X f \, d\mu \right| \leqslant \int_X |f| \, d\mu$.

    d) Show that the integral satisfies the triangle inequality

$$\int_X |f + g| \, d\mu \leqslant \int_X |f| \, d\mu + \int_X |g| \, d\mu$$

and $\int_X |f| \, d\mu = 0$ if and only if $f(x) = 0$ almost everywhere.

    e) Show that $d(f, g) = \int_X |f - g| \, d\mu$ is a semimetric on $\mathfrak{S}_1$.

    f) Show that the integral is countably additive as a set function:

$$\int_{\dot{\bigcup}_{i=1}^{\infty} A_i} f \, d\mu = \sum_{i \in \mathbb{N}} \int_{A_i} f \, d\mu$$

provided that $A = \bigcup_{i \in \mathbb{N}} A_i$ has finite measure and the sets A_i are measurable and mutually disjoint. (Hint: Show that *finite* additivity follows from part (a) above. Show that

$$\int_{\dot{\bigcup}_{i=1}^{\infty} A_i} f \, d\mu = \sum_{i=1}^{N} \int_{A_i} f \, d\mu + \int_{\dot{\bigcup}_{i=N+1}^{\infty} A_i} f \, d\mu,$$

and prove that one of these terms approaches zero as $N \to \infty$.)

5.2.1 The Class $\mathcal{L}^+$ of Nonnegative Measurable Functions

We are ready to take the next-to-last step in the construction of the family of *integrable functions* in the measure space $(X, \mathfrak{A}, \mu)$. Specifically, we will be defining the class $\mathcal{L}^+$ of all *nonnegative* integrable functions on the given abstract measure space. (The letter $\mathcal{L}$ is used in honor of Lebesgue.)

Let f be any extended real-valued, nonnegative, $\mathfrak{A}$-measurable function defined on X. Let $A \in \mathfrak{A}$ and define the *truncation*

$$f_A^M(x) = \begin{cases} f(x) & \text{if } x \in A, \ f(x) \leqslant M, \\ M & \text{if } x \in A, \ f(x) > M, \\ 0 & \text{if } x \in X \backslash A. \end{cases} \tag{5.2}$$

Observe that $f_A^M \in \mathfrak{S}_1$ provided that $\mu(A) < \infty$. Continue to denote $\Phi(f) = \int_X f d\mu$ if $f \in \mathfrak{S}_1$.

Definition 5.2.3 Let f be any extended real-valued, nonnegative, $\mathfrak{A}$-measurable function defined on X, and define the truncation f_A^M as in Equation 5.2. Then we define

$$\int_X f\,d\mu = \Phi(f) = \sup_{\substack{0 < M < \infty \\ A \in \mathfrak{A},\ \mu(A) < \infty}} \Phi\left(f_A^M\right).$$

Definition 5.2.4 An extended real-valued, nonnegative, $\mathfrak{A}$-measurable function f is said to be *integrable* provided that $\Phi(f) < \infty$. The class of all such functions is denoted by $\mathcal{L}^+$.

Lemma 5.2.1 *If the measure space $(X, \mathfrak{A}, \mu)$ is approximately finite, and if $f \in \mathcal{L}^+$, then f has a σ-finite carrier: That is, f vanishes off some σ-finite set.*

Proof: We are given that $\int_X f\,d\mu < \infty$. However, the minimal carrier C of f can be expressed as the set

$$C = \bigcup_{n \in \mathbb{N}} f^{-1}\left(\frac{1}{n}, \infty\right].$$

If any one of these inverse images had infinite measure, then monotonicity together with the approximate finiteness of the measure space would imply that the integral of f must be infinite. This contradiction proves the lemma.   ∎

Lemma 5.2.2 *Assume that f is nonnegative, R^*-valued, and $\mathfrak{A}$-measurable. If $(X, \mathfrak{A}, \mu)$ is σ-finite, then $X = \bigcup_{i \in \mathbb{N}} A_i$, with $\mu(A_i) < \infty$ for each i, and*

$$A_1 \subset A_2 \subset \ldots \subset A_n \subset \ldots.$$

Then the following are equivalent forms of Definition 5.2.3:

i. $\Phi(f) = \lim_{i \to \infty} \lim_{M \to \infty} \Phi(f_{A_i}^M).$

ii. $\Phi(f) = \sup_{\substack{0 \leq g \leq f \\ g \in \mathfrak{S}_1}} \Phi(g).$

iii. $\Phi(f) = \sup_{\substack{0 \leq g \leq f \\ g \in \mathfrak{S}_0}} \Phi(g).$

We leave the proofs of these equivalences to the reader. However, the third form should be noted carefully, because it does not involve the intermediate step of employing $\mathfrak{S}_1$. It is correct because $\Phi(f)$ can be approximated within $\epsilon/2$ by $\Phi(g)$, with $g \in \mathfrak{S}_1$. And $\Phi(g)$ can be approximated within $\epsilon/2$ by $\Phi(h)$, with $h \in \mathfrak{S}_0$ and with $h \leq g \leq f$. See Remark 5.2.1. (An inequality among functions with no variable named means that the inequality is valid for all values of the variable in the domain.)

As for the several preceding domains for the integral, we define

$$\int_A f\,d\mu = \int_X f\,1_A\,d\mu$$

for each $f \in \mathcal{L}^+$ and for each $A \in \mathfrak{A}$.

Theorem 5.2.1 *The integral* $\Phi : \mathcal{L}^+ \to \mathbb{R}$ *has the following properties:*

i. Additivity and Positive Homogeneity: $\Phi(\alpha f + g) = \alpha \Phi(f) + \Phi(g)$ *if* $\alpha \geqslant 0$.

ii. Monotonicity: *If* $f \leqslant g$, *then* $\Phi(f) \leqslant \Phi(g)$, *and in particular,* $0 \leqslant g$ *implies that* $0 < \Phi(g)$.

iii. Triangle Inequality: $\Phi(|f - h|) \leqslant \Phi(|f - g|) + \Phi(|g - h|)$ *if* f, g, *and* h *are in* $\mathcal{L}^+$.

iv. Positive Definiteness:[42] $\Phi(f) = 0$ *if and only if* $f(x) = 0$ *almost everywhere.*

v. Countable Additivity as a Set Function:

$$\int_{\dot{\bigcup}_{i \in \mathbb{N}} A_i} f \, d\mu = \sum_{i \in \mathbb{N}} \int_{A_i} f \, d\mu$$

if the measurable sets A_i *are* mutually disjoint.

Proof:

i. One should note that $\mathcal{L}^+$ is not a vector space. Observe that because Φ is monotone on $\mathfrak{S}_1$, we have

$$\Phi\left((f + g)_A^M\right) \overset{(a)}{\leqslant} \Phi\left(f_A^M + g_A^M\right)$$
$$\overset{(b)}{\leqslant} \Phi\left((f + g)_A^{2M}\right) \qquad (5.3)$$
$$\overset{(c)}{\leqslant} \Phi(f + g)$$

for all A and M. The inequality (a) in Inequalities (5.3), together with the finite additivity of Φ on $\mathfrak{S}_1$, gives us

$$\Phi(f + g) \leqslant \Phi(f) + \Phi(g) < \infty.$$

This proves also that $\mathcal{L}^+$ is closed under addition. Moreover, each individual term labeled with M and with A is monotonically increasing with increasing value of M, and with increasing A by inclusion, because Φ is monotone on $\mathfrak{S}_1$. The inequalities (b) and (c) in Inequalities (5.3) tell us that

$$\Phi\left(f_A^M\right) + \Phi\left(g_{A'}^{M'}\right) \leqslant \Phi\left(f_{A \cup A'}^{\max(M, M')} + g_{A \cup A'}^{\max(M, M')}\right)$$
$$\leqslant \Phi(f + g)$$

---

[42] As in Remark 5.1.1, the property proven here becomes positive definiteness in the proper sense of the word when we form a quotient space $\mathcal{L}/ \sim$, where $f \sim g$ if and only if $f = g$ almost everywhere.

for all A, A', M and M'. Taking the suprema separately over A, M and A', M' yields

$$\Phi(f) + \Phi(g) \leqslant \Phi(f + g),$$

proving additivity. Homogeneity under multiplication by a nonnegative scalar is easy to verify.

ii. Monotonicity is clear since $\Phi(f) \geqslant 0$ if $f \geqslant 0$.

iii. Although $\mathcal{L}^+$ is not a vector space, it is still true that if f and g lie in $\mathcal{L}^+$, then $|f - g| \in \mathcal{L}^+$. The triangle inequality is clear, using (i) and (ii) above, since

$$|f - g| \leqslant |f - h| + |h - g|.$$

iv. Positive definiteness follows from the fact that $\Phi(f) = 0$ if and only if $\Phi\left(f_A^M\right) = 0$ for all A and all M. Since $f_A^M \in \mathfrak{S}_1$, we know that $f_A^M = 0$ almost everywhere. The reader should verify that this implies the same for f itself.

v. Because Φ is finitely additive as a function of the integrand, it is also finitely additive over unions of finitely many *disjoint* domains. In fact, we can write

$$\int_{\dot{\bigcup}_1^N A_i} f \, d\mu = \int_X f \, 1_{\dot{\bigcup}_{i=1}^N A_i} \, d\mu$$

$$= \int_X f \sum_{i=1}^N 1_{A_i} \, d\mu$$

$$= \sum_{i=1}^N \int_{A_i} f \, d\mu.$$

Thus we *could* establish that

$$\int_{\dot{\bigcup}_1^\infty A_i} f \, d\mu = \int_{\dot{\bigcup}_1^N A_i} f \, d\mu + \int_{\dot{\bigcup}_{N+1}^\infty A_i} f \, d\mu$$

$$= \sum_i^N \int_{A_i} f \, d\mu + \int_{\dot{\bigcup}_{N+1}^\infty A_i} f \, d\mu$$

$$\to \sum_i^\infty \int_{A_i} f \, d\mu,$$

which must then equal $\displaystyle\int_{\dot{\bigcup}_1^\infty A_i} f \, d\mu$, *provided* that we can show that

$$\int_{\dot{\bigcup}_{N+1}^\infty A_i} f \, d\mu \to 0$$

as $N \to \infty$.

Let $\epsilon > 0$. There exists a truncation f_A^M such that

$$\int_X f \, d\mu - \int_X f_A^M \, d\mu < \frac{\epsilon}{2}.$$

Therefore

$$\int_{\dot{\bigcup}_{N+1}^{\infty} A_i} f \, d\mu - \int_{\dot{\bigcup}_{N+1}^{\infty} A_i} f_A^M \, d\mu < \frac{\epsilon}{2}$$

for all N. However, we know that $f_A^M \in \mathfrak{S}_1$, so that the integral of f_A^M is a countably additive set function. It follows that

$$\int_{\dot{\bigcup}_{N+1}^{\infty} A_i} f_A^M \, d\mu \to 0$$

as $N \to \infty$. This implies that, for all sufficiently big N, we have

$$\int_{\dot{\bigcup}_{N+1}^{\infty} A_i} f \, d\mu < \epsilon,$$

and this concludes the proof of the theorem.

∎

5.2.2 The Class $\mathcal{L}$ of Lebesgue Integrable Functions

We are ready at last to define the general concept of a Lebesgue integrable function and the Lebesgue integral.

Definition 5.2.5 An $\mathfrak{A}$-measurable, extended real-valued function $f : X \to \mathbb{R}^*$ is said to be integrable provided that *both* $f^+ = \max(f, 0)$ and $f^- = \max(-f, 0)$ belong to $\mathcal{L}^+$. In this case we say that $f \in \mathcal{L}$, the set of all integrable functions, and we define

$$\Phi(f) = \Phi(f^+) - \Phi(f^-).$$

The value of $\Phi(f)$ is normally written as $\int_X f d\mu$.

We remark that if f^+ and f^- are defined as above, then $f = f^+ - f^-$. Also, $|f| = f^+ + f^-$.

EXERCISE

5.6 Let $f : X \to \mathbb{R}^*$ be any $\mathfrak{A}$-measurable function. Prove that $f \in \mathcal{L}$ if and only if $|f| \in \mathcal{L}^+$. Use this to prove that $\mathcal{L}$ is a *vector space*.

Theorem 5.2.2 *Let f and g be in $\mathcal{L}$, and let α be in $\mathbb{R}$.*

i. If $f = f_1 - f_2$ with f_1 and f_2 in $\mathcal{L}^+$, then

$$\int_X f d\mu = \Phi(f_1) - \Phi(f_2).$$

(We do not *assume here that $f_1 = f^+$ or that $f_2 = f^-$.)*

ii. Linearity: $\Phi(\alpha f + g) = \alpha\Phi(f) + \Phi(g)$.

iii. Monotonicity: *If $f \leqslant g$, then $\Phi(f) \leqslant \Phi(g)$. Also,* $\int_X |f|\, d\mu = 0$ *if and only if $f = 0$ almost everywhere.*

iv. Triangle Inequality: *If $f \in \mathcal{L}$, then $|f| \in \mathcal{L}$ and $\left|\int_X f\, d\mu\right| \leqslant \int_X |f|\, d\mu$.*

v. Countable Additivity as a Set Function: *Let the integral over a measurable subset $A \subseteq X$ be defined by*

$$\int_A f\, d\mu = \int_X f \cdot 1_A\, d\mu.$$

Then the integral over A is countably additive as a set function of A.

Proof:

i. In this case, $f = f_1 - f_2 = f^+ - f^-$ expresses f in two ways as the difference of functions in $\mathcal{L}^+$. Hence $f_1 + f^- = f^+ + f_2$. Thus

$$\Phi(f_1 + f^-) = \Phi(f_1) + \Phi(f^-) = \Phi(f^+) + \Phi(f_2).$$

It follows that

$$\int_X f\, d\mu = \Phi(f^+) - \Phi(f^-) = \Phi(f_1) - \Phi(f_2).$$

ii. If $f = f_1 - f_2$, as in the preceding part, and if $\alpha > 0$, then $\Phi(\alpha f) = \alpha\Phi(f)$ by Theorem 5.2.1. And if $\alpha < 0$, then $\alpha f = |\alpha|(f_2 - f_1)$, so that

$$\begin{aligned}
\Phi(\alpha f) &= |\alpha|\Phi(f_2) - |\alpha|\Phi(f_1) \\
&= \alpha[\Phi(f_1) - \Phi(f_2)] \\
&= \alpha\Phi(f).
\end{aligned}$$

Now suppose also that $g = g_1 - g_2$ with g_1 and g_2 in $\mathcal{L}^+$. Then

$$\Phi(f + g) = \Phi(f_1 + g_1) - \Phi(f_2 + g_2) = \Phi(f) + \Phi(g).$$

iii. If $f \leqslant g$, then $g = f + (g-f)$ and $\Phi(g) = \Phi(f) + \Phi(g-f)$, where $\Phi(g-f) \geqslant 0$ since $g - f \in \mathcal{L}^+$. Also, if $L(|f|) = 0$, then $|f| = 0$ almost everywhere, and the same is true for f. Conversely, if $f = 0$ almost everywhere, then $L(|f|) = 0$.

iv. Note that

$$\begin{aligned}
\left|\int_X f\, d\mu\right| &= \left|\int_X f^+\, d\mu - \int_X f^-\, d\mu\right| \\
&\leqslant \int_X f^+ + f^-\, d\mu \\
&= \int_X |f|\, d\mu.
\end{aligned}$$

v. We write $f = f^+ - f^-$ as before and let $A_i \in \mathcal{A}$ be a sequence of mutually disjoint sets. Then

$$\int_{\dot{\bigcup}_{i \in \mathbb{N}} A_i} f^+ \, d\mu = \sum_{i \in \mathbb{N}} \int_{A_i} f^+ \, d\mu.$$

The latter series is *absolutely convergent*, being bounded by $\int_X |f| \, d\mu$, as in the preceding part. We know this from an earlier theorem establishing the countable additivity of this function for functions in $\mathcal{L}^+$.

Also,

$$\int_{\dot{\bigcup}_{i \in \mathbb{N}} A_i} f^- \, d\mu = \sum_{i \in \mathbb{N}} \int_{A_i} f^- \, d\mu,$$

with the series on the right side being absolutely convergent. It follows from standard theorems about absolutely convergent series that

$$\sum_{i \in \mathbb{N}} \int_{A_i} f \, d\mu = \sum_{i \in \mathbb{N}} \int_{A_i} (f^+ - f^-) \, d\mu$$

$$= \sum_{i \in \mathbb{N}} \int_{A_i} f^+ \, d\mu - \sum_{i \in \mathbb{N}} \int_{A_i} f^- \, d\mu$$

$$= \int_{\dot{\bigcup}_{i \in \mathbb{N}} A_i} f^+ \, d\mu - \int_{\dot{\bigcup}_{i \in \mathbb{N}} A_i} f^- \, d\mu$$

$$= \int_{\dot{\bigcup}_{i \in \mathbb{N}} A_i} f \, d\mu.$$

$\blacksquare$

EXERCISES

5.7 Prove that the Lebesgue integral on $\mathbb{R}^n$ is translation-invariant. That is, if $f \in \mathcal{L}(\mathbb{R}^n)$ and if $f_t(x) = f(x + t)$ for all x and t in $\mathbb{R}^n$, then

$$\int_{\mathbb{R}^n} f \, dl = \int_{\mathbb{R}^n} f_t \, dl.$$

5.8 Prove that the Lebesgue integral on $\mathbb{R}^n$ is invariant under reflections through the origin. That is, if $f \in \mathcal{L}(\mathbb{R}^n)$, then

$$\int_{\mathbb{R}^n} f(-x) \, dl(x) = \int_{\mathbb{R}^n} f(x) \, dl(x).$$

5.9 Define the σ-finite *counting measure* ν in the measure space $(\mathbb{N}, \mathfrak{P}(\mathbb{N}), \nu)$ as in Exercise 2.15. Show that a function $f : \mathbb{N} \to \mathbb{R}$ is *integrable* on $\mathbb{N}$ if and only if the sequence $f(n)$ is absolutely summable. Prove that if f is integrable, then

$$\int_{\mathbb{N}} f \, d\nu = \sum_{n \in \mathbb{N}} f(n).$$

5.10 Using Definition 1.2.2, show that if $f : [a, b] \to \mathbb{R}$ is a Riemann integrable function, then it must be Lebesgue integrable as well, and the Lebesgue integral has the same value as the Riemann integral of f. (Hint: Each Riemann integrable function can be approximated from below by step functions, which are *very* special simple functions.)

5.2.3 Convex Functions and Jensen's Inequality

In this subsection, we will prove a useful generalization, known as *Jensen's inequality*, of the triangle inequality from Theorem 5.2.2. The reader should recall from advanced calculus that a connected subset of the real line is called an *interval*, whether it is finitely long or infinitely long. A subset S of a vector space is called *convex* provided that for each pair of points P and Q in S, the straight line segment joining P to Q lies in S.

Definition 5.2.6 We call a function $\phi : I \to \mathbb{R}$ a *convex function* on the interval I if and only if the so-called *epigraph*,

$$\varepsilon(\phi) = \{(t, y) \mid y \geqslant \phi(t)\},$$

is a *convex* subset of the plane. A function ϕ is called *concave* provided that $-\phi$ is convex.

The following lemma identifies an important geometrical property of convex functions.

Lemma 5.2.3 *Let $\phi : I \to \mathbb{R}$ be a convex function on the interval I, and let c be any interior point of I. Then there exists a real number m such that*

$$m(t - c) + \phi(c) \leqslant \phi(t), \tag{5.4}$$

for all $t \in I$. A straight line that is the graph in the plane of $y = m(t - c) + \phi(c)$, satisfying Equation (5.4), is called a supporting line *for ϕ at $t = c$.*

Proof: We let

$$C^+ = \left\{ \frac{\phi(t) - \phi(c)}{t - c} \,\middle|\, t > c \right\} \quad \text{and} \quad C^- = \left\{ \frac{\phi(t) - \phi(c)}{t - c} \,\middle|\, t < c \right\},$$

with t required to belong to I. The convexity of $\varepsilon(\phi)$ implies that each element of C^- is a lower bound of C^+: Otherwise, the convexity of $\varepsilon(\phi)$ would produce an element of the epigraph below the point $(c, \phi(c))$. The reader should do a calculation with convex combinations of points to verify this claim by letting $t_1 < c < t_2$ and showing that

$$\frac{\phi(t_1) - \phi(c)}{t_1 - c} \leqslant \frac{\phi(t_2) - \phi(c)}{t_2 - c}.$$

Hence there is a real number

$$m = \inf(C^+).$$

One can check readily that m satisfies Inequality (5.4).   ∎

Jensen's inequality pertains to integrals on *probability spaces*, which we define below.

Definition 5.2.7 A measure space $(X, \mathfrak{A}, \mu)$ is called a *probability space*, and the nonnegative measure μ is called a *probability measure*, provided that $\mu(X) = 1$.

Theorem 5.2.3 (Jensen's Inequality) *Let ϕ be any convex, Borel measurable function defined an interval I, which may be either finite or infinite. Let f be a real-valued integrable function on a probability space $(X, \mathfrak{A}, \mu)$. Suppose that the range of f is contained in I. Then*

$$\phi\left(\int_X f \, d\mu\right) \leqslant \int_X \phi \circ f \, d\mu,$$

provided that $\phi \circ f$ is integrable.

Proof: We begin by letting $c = \displaystyle\int_X f \, d\mu$. If c were an endpoint of I, it would follow that the function f is equal almost everywhere to a constant, and this would imply that Jensen's inequality is satisfied by being an equality. The main case is that in which c is an interior point of I, and then we let m denote the slope of any supporting line $y = m(t - c) + \phi(c)$ at $t = c$ for the convex function ϕ. Thus

$$m(t - c) + \phi(c) \leqslant \phi(t)$$

for all $t \in I$. It follows from monotonicity of the integral that

$$\begin{aligned}
\int_X \phi \circ f \, d\mu &\geqslant \int_X m\big(f(x) - c\big) + \phi(c) \, d\mu(x) \\
&= m\left(\int_X f \, d\mu - c\right) + \int_X \phi(c) \, d\mu \\
&= \phi(c) \\
&= \phi\left(\int_X f \, d\mu\right),
\end{aligned}$$

since μ is a probability measure.   ∎

Remark 5.2.2 Observe that since ϕ is Borel measurable, it follows that $\phi \circ f$ is μ-measurable. (See Exercise 5.14.) The connection between Jensen's inequality and the triangle inequality, from Theorem 5.2.2, is that the absolute value function is convex. It is possible to strengthen the statement of Jensen's inequality slightly by allowing for the possibility that $\int_X \phi \circ f \, d\mu = \infty$, although in that case $\phi \circ f$ is not integrable. That is, it can occur that the positive part, $(\phi \circ f)^+$, fails to be integrable, although the negative part is integrable, because of the existence of a supporting line for the convex function ϕ. (See Exercise 5.15.)

Also, the reader should note that Jensen's inequality implies easily that if ϕ were a *concave* Borel function that is defined on an interval containing the range of f, then Jensen's inequality would imply that

$$\phi \left(\int_X f \, d\mu \right) \geq \int_X \phi \circ f \, d\mu.$$

EXERCISES

5.11   Suppose f is a nonnegative integrable function on $[0, 1]$ with respect to Lebesgue measure. Prove that

$$\sqrt{\int_0^1 f \, dl} \geq \int_0^1 \sqrt{f} \, dl.$$

5.12   Let f be a nonnegative integrable function on $[0, 1]$ and let $I = \int_0^1 f \, dl$, where l is Lebesgue measure. Show that

$$\sqrt{1 + I^2} \leq \int_0^1 \sqrt{1 + f^2} \, dl \leq 1 + I.$$

5.13   Let $(X, \mathfrak{A}, \mu)$ be a probability space. Suppose that f is a measurable function with $\|f\|_{\sup} < \infty$ and that $1 \leq p < q < \infty$ with p and q being real numbers. Prove that[43]

$$\left(\int_X |f|^p \, d\mu \right)^{\frac{1}{p}} \leq \left(\int_X |f|^q \, d\mu \right)^{\frac{1}{q}}.$$

5.14   Let ϕ be a convex function defined on an interval I. Prove that ϕ is continuous on the interior of I, thereby establishing that ϕ is Borel measurable. (Hint: At each c in I°, use a supporting line for $\varepsilon(\phi)$ and a chord to prove continuity.)

5.15   Give an example of a convex function ϕ on an interval of the real line, together with an integrable function f such that $\phi \circ f$ is not integrable. Show that the negative part, $(\phi \circ f)^-$, must be integrable.

5.3   LEBESGUE DOMINATED CONVERGENCE THEOREM

The Lebesgue integral was developed especially to deal with limits of integrals and integrals of limits. The most famous theorem about this topic is the Lebesgue

---

[43]For a somewhat strengthened form of this exercise, see Exercise 9.2.

Dominated Convergence theorem. (This theorem is so widely used, that it is often cited simply as LDC.

Theorem 5.3.1 (Lebesgue Dominated Convergence) *Let $(X, \mathfrak{A}, \mu)$ be a complete measure space. Let $f_n : X \to \mathbb{R}^*$ be a sequence of $\mathfrak{A}$-measurable functions converging pointwise almost everywhere to a function f. Suppose there exists an integrable function $\phi \in \mathcal{L}$ such that $|f_n(x)| \leqslant \phi(x)$ for all $x \in X$ and for all $n \in \mathbb{N}$. Then*

   i. f is integrable and

   ii. $\displaystyle\int_X f\,d\mu = \lim_{n \to \infty} \int_X f_n\,d\mu.$

Remark 5.3.1 The conclusion of this theorem can be rewritten slightly in the form

$$\int_X \lim_{n \to \infty} f_n\,d\mu = \lim_{n \to \infty} \int_X f_n\,d\mu.$$

This form emphasizes that the Lebesgue Dominated Convergence theorem establishes a sufficient condition for interchanging the order of two very important limits: the integral, which is a very intricate limit, and the pointwise limit of a sequence of functions. Many of the most important and useful theorems in analysis are concerned with such an interchange in the order of taking limits. Example 1.3 is an illustration that the successive application of limit operations need not be commutative.

    Note that the existence of the *dominating function*, $\phi \in \mathcal{L}$, ensures that f is finite almost everywhere.

Proof: Observe first that because

$$|f_n| = f_n^+ + f_n^- \leqslant \phi,$$

it follows that both f_n^+ and f_n^- lie in $\mathcal{L}^+$ and thus each $f_n \in \mathcal{L}$. Thus we could as well have assumed that each f_n is integrable. Also, by Theorem 4.2.1, we know that f is integrable as well.

    What we are observing here is that f is equal almost everywhere to a measurable function. In a complete measure space, this proves that f is measurable.[44]

   i. Suppose that $\mu(X) < \infty$ and that $\phi(x) \equiv M$, a nonnegative real constant. Let $\epsilon > 0$. By Egoroff's theorem, there exists a set $A \in \mathfrak{A}$ such that

$$\mu(A) < \frac{\epsilon}{4M}$$

     and such that $f_n \to f$ *uniformly* on $X \backslash A$. Thus there exists $N \in \mathbb{N}$ such that for all $n \geqslant N$ we have

$$|f_n(x) - f(x)| < \frac{\epsilon}{2\mu(X)}$$

---

[44]In an incomplete measure space, we could require that $f_n(x) \to f(x)$ everywhere on X. Alternatively, we could set $f(x) \equiv 0$ on the Lebesgue null set of nonconvergence.

for all $x \in X \backslash A$. Hence for all $n \geq N$ we have

$$\left| \int_X f_n \, d\mu - \int_X f \, d\mu \right| \leq \int_X |f_n - f| \, d\mu$$

$$= \int_{X \backslash A} |f_n - f| \, d\mu + \int_A |f_n - f| \, d\mu$$

$$< \frac{\epsilon}{2\mu(X)} \mu(X \backslash A) + 2M\mu(A) < \epsilon.$$

It follows that $\int_X f_n \, d\mu \to \int_X f \, d\mu$ as $n \to \infty$.

ii. This will be the general case, and we will see how the first case facilitates the second. Since ϕ is nonnegative and integrable, there exists a special simple function, $\phi_0 \in \mathfrak{S}_0$, such that $0 \leq \phi_0 \leq \phi$ and

$$\int_X \phi_0 \, d\mu > \int_X \phi \, d\mu - \frac{\epsilon}{3}.$$

Since

$$f_n = f_n^+ - f_n^- \to f = f^+ - f^-$$

almost everywhere on X, it follows that $f_n^+ = f_n \vee 0 \to f \vee 0 = f^+$ almost everywhere. Similarly, $f_n^- \to f^-$ almost everywhere. Because f^+ and f^- are nonnegative, it will suffice to prove the theorem for $f \geq 0$. Also,

$$f_n \wedge \phi_0 = \min(f_n, \phi_0) \to f \wedge \phi_0$$

almost everywhere on X, and $0 \leq f_n \wedge \phi_0 \leq \phi_0$.

Because special simple functions are finite linear combinations of indicator functions of measurable sets of finite measure, the proof in case (i) applies as well to upper bounds that are special simple functions as it applies to upper bounds that are constant on a set of finite measure. Hence case (i) tells us that there is an $N \in \mathbb{N}$ such that if $n \geq N$, we have

$$\int_X |f_n \wedge \phi_0 - f \wedge \phi_0| \, d\mu < \frac{\epsilon}{3}.$$

Next, observe that since $0 \leq \phi_0 \wedge f_n \leq f_n \leq \phi$, it follows that

$$\int_X f_n - f_n \wedge \phi_0 \, d\mu \leq \int_X \phi - \phi_0 \, d\mu < \frac{\epsilon}{3}$$

and

$$\int_X f - f \wedge \phi_0 \, d\mu \leq \int_X \phi - \phi_0 \, d\mu < \frac{\epsilon}{3}.$$

Thus for each $n \geq N$ we have

$$\left| \int_X f \, d\mu - \int_X f_n \, d\mu \right| \leq \left| \int_X f \, d\mu - \int_X f \wedge \phi_0 \, d\mu \right|$$
$$+ \left| \int_X f \wedge \phi_0 \, d\mu - \int_X f_n \wedge \phi_0 \, d\mu \right|$$
$$+ \left| \int_X f_n \wedge \phi_0 \, d\mu - \int_X f_n \, d\mu \right|$$
$$< \epsilon.$$

Thus we have proven that

$$\lim_{n \to \infty} \int_X f_n \, d\mu = \int_X f \, d\mu.$$

■

■ EXAMPLE 5.1

Let $S = \{r_n \mid n \in \mathbb{N}\} = \mathbb{Q} \cap [0,1]$. Let $f_n = 1_{\{r_1,\ldots,r_n\}}$ for each $n \in \mathbb{N}$. Thus $|f_n| \leq 1 \in \mathcal{L}[0,1]$ for all n and $f_n \to 1_S$ as $n \to \infty$. This implies by the Lebesgue Dominated Convergence theorem that $1_S \in \mathcal{L}[0,1]$ and that

$$\int_{[0,1]} 1_S \, d\mu = \lim_{n \to \infty} \int_{[0,1]} f_n \, d\mu = 0.$$

However, 1_S is not even Riemann integrable on $[0,1]$.

EXERCISES

5.16   Let the graph of $f_n : [0,1] \to \mathbb{R}$ be an *isosceles* triangle with *base* $\left[\frac{1}{n+1}, \frac{1}{n}\right]$ and *altitude* a_n. Prove that $f_n \to f \equiv 0$ on $[0,1]$, but that

$$\int_{[0,1]} f_n \, d\mu \to \infty$$

as $n \to \infty$, provided that the sequence a_n grows sufficiently rapidly as n increases. Thus domination (boundedness) by an integrable function is a necessary hypothesis in Theorem 5.3.1.

5.17   Let

$$f_n(x) = \frac{x}{n} 1_{[-n,n]}(x).$$

Find the pointwise limit $f(x) = \lim_{n \to \infty} f_n(x)$ on $\mathbb{R}$. Prove that

$$\int_{\mathbb{R}} f_n \, d\mu \to \int_{\mathbb{R}} f \, d\mu.$$

Does the sequence f_n satisfy the hypotheses of the Lebesgue Dominated Convergence theorem? Explain.

5.18   With respect to Lebesgue measure l on $[0, 1]$, give an example of a *nonmeasurable* function $f : [0, 1] \to \mathbb{R}$ such that $|f| \in \mathcal{L}[0, 1]$ and $\int_{[0,1]} |f| \, dl = 1$.

5.19   Let f be integrable on an *arbitrary* measure space $(X, \mathfrak{A}, \mu)$. Let $A_n \in \mathfrak{A}$ be an arbitrary sequence of *disjoint* measurable sets. In Theorem 5.2.2 we proved that

$$\int_{\dot{\bigcup}_{n \in \mathbb{N}} A_n} f \, d\mu = \sum_{n \in \mathbb{N}} \int_{A_n} f \, d\mu.$$

Give an alternative proof using the Lebesgue Dominated Convergence theorem.

5.20   Let $f \in \mathcal{L}[1, \infty)$ with respect to Lebesgue measure. Prove or disprove:
a) $\int_{[1,b]} f \, dl \to 0$ as $b \to 1+$.
b) $\int_{[b,\infty)} f \, dl \to 0$ as $b \to \infty$.
c) This is a generalization of parts (a) and (b) above. Suppose that

$$f \in \mathcal{L}(X, \mathfrak{A}, \mu)$$

is integrable on a general measure space. Prove that[45]
   i. If $\epsilon > 0$, there exists a set $A \in \mathfrak{A}$ of finite measure such that

$$\int_{A^c} |f| \, d\mu < \epsilon.$$

   ii. $\int_A f \, d\mu \to 0$ as $\mu(A) \to 0$, where $A \in \mathfrak{A}$. (Hint: Use the definition of $\mathcal{L}(X, \mathfrak{A}, \mu)$ in terms of bounded nonnegative measurable functions with finite carrier.)

5.21   Let

$$f_n(x) = \begin{cases} n - n|x| & \text{if } |x| \leq \frac{1}{n} \\ 0 & \text{if } |x| > \frac{1}{n}. \end{cases}$$

Prove that $f_n \to 0$ pointwise almost everywhere. Is

$$\lim_{n \to \infty} \int_{\mathbb{R}} f_n \, dl = \int_{\mathbb{R}} \lim_{n \to \infty} f_n \, dl?$$

Reconcile your conclusions with the Lebesgue Dominated Convergence theorem.

5.22   Let $f \in \mathcal{L}(\mathbb{R})$. Find

$$\lim_{n \to \infty} \int_{\mathbb{R}} e^{-nx^2} f(x) \, dl(x)$$

and justify your conclusion.

---

[45] Variations on this exercise appear in Exercises 5.24 and 8.10. Compare also with Exercise 2.19.

5.23   Let $f_n : [0, \infty) \to \mathbb{R}$ be defined by

$$f_n(x) = \frac{x}{n} e^{-\frac{x}{n}}$$

for each $n \in \mathbb{N}$. Let $f(x) = \lim_{n \to \infty} f_n(x)$. Show that for all $a \in [0, \infty)$ we have

$$\lim_{n \to \infty} \int_0^a f_n(x) \, dx = \int_0^a f(x) \, dx$$

but that

$$\lim_{n \to \infty} \int_0^\infty f_n(x) \, dx \neq \int_0^\infty f(x) \, dx.$$

Explain.

5.24   Let $E_1 \supset E_2 \supset \ldots \supset E_n \supset \ldots$ be a decreasing nest of measurable sets in the complete measure space $(X, \mathfrak{A}, \mu)$. Let f be integrable on $(X, \mathfrak{A}, \mu)$ and suppose that

$$\mu \left(\bigcap_{n \in \mathbb{N}} E_n \right) = 0.$$

Prove[46] that $\int_{E_n} f \, d\mu \to 0$ as $n \to \infty$.

5.25   Let $f \in \mathcal{L}(\mathbb{R})$, and suppose also that $\int_{\mathbb{R}} |x f(x)| \, dl(x) < \infty$. Define

$$F(\alpha) = \int_{\mathbb{R}} f(x) \cos(\alpha x) \, dl(x)$$

for all $\alpha \in \mathbb{R}$. Prove that the *derivative*

$$F'(\alpha) = \frac{d}{d\alpha} \int_{\mathbb{R}} f(x) \cos(\alpha x) \, dl(x)$$

exists and find its value for all $\alpha \in \mathbb{R}$. Observe that this problem deals with the interchange of order of two limits: the integral and the derivative. You will need to *justify carefully* bringing the derivative with respect to α inside the integral. (Hints: Take a sequence $h_n \to 0$ and show that

$$\lim_{n \to \infty} \frac{F(\alpha + h_n) - F(\alpha)}{h_n}$$

exists and is independent of the choice of $h_n \to 0$.)

5.26   Let $a_{m,n}$ be a double sequence of real numbers having the property that there exists a sequence b_m for which $|a_{m,n}| \leqslant |b_m|$ for all m and n in $\mathbb{N}$ and such that $\sum_m |b_m| < \infty$. Suppose also that $a_{m,n} \to A_m$, as $n \to \infty$ for each m.
    a) Prove that

$$\sum_m A_m = \lim_n \sum_m a_{m,n}.$$

---

[46] Another proof is suggested in Exercise 8.10.

(Hint: Apply Lebesgue Dominated Convergence, interpreting the summation on m as being an integral over $\mathbb{N}$ with respect to counting measure.)

b) Give an example in which no summable sequence such as b_m exists, in which the conclusion of part (a) fails.

5.4 MONOTONE CONVERGENCE AND FATOU'S THEOREM

The following theorem extends somewhat the conclusions of the Lebesgue Dominated Convergence theorem in the case of monotone increasing sequences of nonnegative, measurable functions.

Theorem 5.4.1 (Monotone Convergence) *Let* $(X, \mathfrak{A}, \mu)$ *be a complete measure space, and suppose* $f_n : X \to \mathbb{R}$ *is* $\mathfrak{A}$-*measurable for each* $n \in \mathbb{N}$. *If each* $f_n \geqslant 0$ *and if the sequence* $f_n(x)$ *is increasing monotonically to the limit* $f(x)$ *almost everywhere, then*

$$\int_X \lim_{n \to \infty} f_n \, d\mu = \lim_{n \to \infty} \int_X f_n \, d\mu.$$

Before proving the theorem, we observe that the Monotone Convergence theorem differs from the Lebesgue Dominated Convergence theorem in that we do not assume that the sequence f_n is bounded above by an integrable function. Note that $f(x)$ may be infinite. For example, the Monotone Convergence theorem will still apply even if $f_n(x)$ diverges to infinity at each x, though it would assert in that case only that both sides of the equation are infinite.

Proof: We know that f is measurable because it is the pointwise limit almost everywhere of measurable functions, combined with the hypothesis that the measure space is complete. If $f \in \mathcal{L}(X, \mathfrak{A}, \mu)$, then the Monotone Convergence theorem is an immediate consequence of the Lebesgue Dominated Convergence theorem. Beyond the content of the Lebesgue Dominated Convergence theorem, the Monotone Convergence theorem adds only the claim that if $\int_X f \, d\mu = \infty$ then $\int_X f_n \, d\mu \to \infty$ too.

Suppose therefore that $\int_X f \, d\mu = \infty$. This means that for each $M > 0$ there exists $\phi \in \mathfrak{S}_0$ such that $0 \leqslant \phi \leqslant f$ and $M < \int_X \phi \, d\mu < \infty$. Also, $f_n \wedge \phi(x)$ is an increasing nonnegative sequence converging almost everywhere to the dominating function $f \wedge \phi(x)$. Thus the Lebesgue Dominated Convergence theorem implies that

$$\int_X f_n \wedge \phi \, d\mu \to \int_X f \wedge \phi \, d\mu = \int_X \phi \, d\mu > M.$$

Hence there exists N such that $n \geqslant N$ implies that

$$\int_X f_n \, d\mu \geqslant \int_X f_n \wedge \phi \, d\mu > M.$$

This proves the theorem.   ∎

As a handy application of the Monotone Convergence theorem, we prove the following result.

Theorem 5.4.2 *Suppose f is measurable on $(X, \mathfrak{A}, \mu)$, a σ-finite measure space. If*

$$X = \bigcup_{n \in \mathbb{N}} A_n$$

is a decomposition into the union of an expanding nest of measurable sets of finite measure, then

$$f \in \mathcal{L}(X, \mathfrak{A}, \mu) \Leftrightarrow \sup \left\{ \int_{A_n} f \, d\mu \,\middle|\, n \in \mathbb{N} \right\} < \infty.$$

Proof: Let $f_n = |f| 1_{A_n}$ for each $n \in \mathbb{N}$. Then f_n is an increasing sequence of nonnegative measurable functions converging everywhere to f. The Monotone Convergence theorem gives the conclusion that $\int_X |f| \, d\mu < \infty$ if and only if $\lim_n \int_X f_n \, d\mu < \infty$.   ∎

A simple application of this theorem would be to show that $f(x) = \dfrac{1}{1 + x^2}$ is integrable on $\mathbb{R}$ by showing that $\lim_n \displaystyle\int_{-n}^{n} \dfrac{1}{1 + x^2} \, dl = \pi < \infty$.

EXERCISES

5.27 Let $f(x) = \frac{1}{\sqrt{x}}$ on the interval $(0, 1)$.
  a) Show that $f \in \mathcal{L}((0, 1), \mathcal{L}, l)$.
  b) If $g \in \mathfrak{S}_0$ is a special simple function satisfying $0 \leqslant g \leqslant f$ on $(0, 1)$, prove that $\int_{(0,1)} g \, dl \leqslant 2$.
  c) Show that $\{h \in \mathfrak{S} \mid f \leqslant h\} = \varnothing$. This helps to explain why the integral of a nonnegative measurable function is defined in terms of approximations from *below* by special simple functions—*not from above*.

5.28 Show first that $g(x) = e^{-|x|}$ is a Lebesgue integrable function on the real line.
  a) Suppose that $|f_n(x)| \leqslant e^{-|x|}$ for each $x \in \mathbb{R}$ and each $n \in \mathbb{N}$. Suppose also that $\int_{\mathbb{R}} f_n \, dl = 0$ for each $n \in \mathbb{N}$ and that $f_n \to f$ pointwise almost everywhere on $\mathbb{R}$. Prove that $f \in \mathcal{L}(\mathbb{R})$ and that $\int_{\mathbb{R}} f \, dl = 0$.
  b) Now let

$$f_n(x) = e^{-|x|} 1_{[-n,n]}(x) \sin x.$$

  Show that

$$f(x) = e^{-|x|} \sin x$$

  is Lebesgue integrable on the real line and that $\int_{\mathbb{R}} f \, dl = 0$.

c) Show that $h(x) = \sin x$ is not a Lebesgue integrable function on $\mathbb{R}$, although

$$\int_{[-n,n]} h \, dl = 0$$

for all $n \in \mathbb{N}$.

5.29 Suppose that $f_n \in \mathcal{L}[0,1]$ for each $n \in \mathbb{N}$ and suppose also that

$$\sum_{n\in\mathbb{N}} \int_0^1 |f_n| \, dl < \infty.$$

Prove that $\sum_{n\in\mathbb{N}} f_n(x)$ is (absolutely) convergent almost everywhere to a function $f \in \mathcal{L}[0,1]$, and prove that

$$\int_0^1 f \, dl = \sum_{n\in\mathbb{N}} \int_0^1 f_n \, dl.$$

5.30 Let $f \in \mathcal{L}(\mathbb{R})$ be a nonnegative function. Suppose that the sequence

$$\int_{\mathbb{R}} f^n \, dl$$

converges to a real number. Prove that the sequence of powers f^n converges almost everywhere to the indicator function of a measurable set.

5.31 Let f be a nonnegative measurable function on a σ-finite measure space, $(X, \mathfrak{A}, \mu)$. Prove that f is the pointwise limit of a *monotone increasing* sequence of special simple functions.

5.32 Find

$$\lim_{n\to\infty} \int_0^1 \sin\left(\frac{x}{n}\right) \frac{n^3}{1 + n^2 x^3} \, dx.$$

Another useful convergence theorem, called either Fatou's Lemma or Fatou's theorem, is a consequence of the Monotone Convergence theorem.

Theorem 5.4.3 (Fatou) *Suppose* $(X, \mathfrak{A}, \mu)$ *is a complete measure space and suppose that* $f_n : X \to \mathbb{R}$ *is* $\mathfrak{A}$-measurable and nonnegative for each $n \in \mathbb{N}$. *Suppose that* $\liminf_n f_n(x) = f(x)$ *almost everywhere. Then*

$$\int_X f \, d\mu \leq \liminf_{n\to\infty} \int_X f_n \, d\mu.$$

Proof: By Theorem 4.2.1 we know that f is measurable. Moreover,

$$\psi_n = \inf\{f_n, f_{n+1}, \ldots\}$$

is an increasing sequence of measurable functions converging pointwise almost everywhere to f. Since ψ_n is monotone increasing, the Monotone Convergence theorem tells us that

$$\int_X \psi_n \, d\mu \to \int_X f \, d\mu.$$

Since $f_k \geqslant \psi_n$ for all $k \geqslant n$, it follows that

$$\inf \left\{ \int_X f_k \, d\mu \,\middle|\, k \geqslant n \right\} \geqslant \int_X \psi_n \, d\mu,$$

which implies in turn that

$$\liminf_{n \to \infty} \int_X f_n \, d\mu \geqslant \lim_{n \to \infty} \int_X \psi_n \, d\mu$$

$$= \int_X f \, d\mu.$$

■

EXERCISES

5.33   Prove that in Fatou's theorem, we can relax the hypothesis that $f_n \geqslant 0$ as follows. Suppose there exists $\phi \in \mathcal{L}^+$ such that $-\phi \leqslant f_n$. Show that the conclusion of Fatou's theorem still follows.

5.34   Give an example of a sequence of nonnegative real-valued $\mathfrak{A}$-measurable functions f_n on $[0, 1]$ such that $f_n \to f$ almost everywhere, yet

$$\liminf_{n \to \infty} \int_{[0,1]} f_n \, dl \neq \int_{[0,1]} f \, dl.$$

5.35   Give an example of a sequence of functions $f_n \in \mathcal{L}(\mathbb{R})$ for which

$$\int_{\mathbb{R}} \liminf_{n \to \infty} f \, d\mu > \liminf_{n \to \infty} \int_{\mathbb{R}} f_n \, d\mu.$$

Explain why your example does not violate Fatou's theorem.

5.5   COMPLETENESS OF $L^1(X, \mathfrak{A}, \mu)$ AND THE POINTWISE CONVERGENCE LEMMA

We saw in Definition 5.2.5, together with Exercise 5.6, that $\mathcal{L}(X, \mathfrak{A}, \mu)$ is a vector space. In this section we will place the structure of a *normed vector space* (also called a *normed linear space*) on $\mathcal{L}(X, \mathfrak{A}, \mu)$. First, we remind the reader what this means.

Definition 5.5.1 A *(real) normed vector space* is a real vector space V, equipped with a real-valued function called a *norm*, denoted by $\| \cdot \|$, provided that for all $\vec{v}$ and $\vec{w}$ in V, and for all $\alpha \in \mathbb{R}$, the following properties are satisfied:

  i. $\|\vec{v}\| \geq 0$, and $\|\vec{v}\| = 0$ if and only if $\vec{v} \equiv \vec{0}$.

  ii. $\|\alpha \vec{v}\| = |\alpha| \|\vec{v}\|$.

  iii. $\|\vec{v} + \vec{w}\| \leq \|\vec{v}\| + \|\vec{w}\|$. (This is the *triangle inequality*.)

The norm that we will use is called the L^1-*norm*, named after Henri Lebesgue.

Definition 5.5.2 Let $(X, \mathfrak{A}, \mu)$ be a measure space. We define the L^1-norm of an integrable function f by

$$\|f\|_1 = \int_X |f| \, d\mu,$$

and we note that for each $f \in \mathcal{L}(X, \mathfrak{A}, \mu)$ we have $\|f\|_1 < \infty$.

EXERCISE

5.36 Show that the L^1-norm on the vector space $\mathcal{L}(\mathbb{R}, \mathfrak{L}, l)$ satisfies properties (ii) and (iii) of Definition 5.5.1 but does *not* satisfy property (i).

The difficulty is that an integrable function f can satisfy $\|f\|_1 = 0$ without f being the (identically) zero function. We can remedy this defect, and turn $\mathcal{L}(X, \mathfrak{A}, \mu)$ into a normed vector space $L^1(X, \mathfrak{A}, \mu)$, as follows.

Definition 5.5.3 We call two functions f and g in $\mathcal{L}(X, \mathfrak{A}, \mu)$ *equivalent*, denoted by $f \sim g$, provided that $\|f - g\|_1 = 0$. We define

$$L^1(X, \mathfrak{A}, \mu) = \mathcal{L}(X, \mathfrak{A}, \mu) / \sim$$
$$= \mathcal{L}(X, \mathfrak{A}, \mu) \Big/ \left\{ f \in \mathcal{L}(X, \mathfrak{A}, \mu) \,\big|\, \|f\|_1 = 0 \right\},$$

the quotient space of the vector space $\mathcal{L}(X, \mathfrak{A}, \mu)$ modulo the subspace of all functions having L^1-norm equal to zero. The equivalence class of an integrable function f is commonly denoted $[f]$.

EXERCISE

5.37 If $f \in \mathcal{L}(X, \mathfrak{A}, \mu)$, prove that $\|f\|_1 = 0$ if and only if $f(x) = 0$ almost everywhere.

When dealing with $L^1(X, \mathfrak{A}, \mu)$, it is very common to go back and forth between considering functions and the equivalence classes of functions. It is not common to denote the integral of an equivalence class, for example. It is easy to check that if $g \in [f]$, then $\int_X f \, d\mu = \int_X g \, d\mu$, so that it does not matter which member of an equivalence class one integrates.

Subtleties arise, however, of the following kind. Sometimes one speaks of some function $f \in L^1(\mathbb{R})$ being a continuous function. What this actually means is that there *exists* a continuous function which can be selected for use as a representative of that equivalence class. That is, there exists a continuous function $g \in [f]$. But other representatives of $[f]$ need not be continuous even though $g \in C(\mathbb{R}) \cap [f]$. Still, if f is continuous, then it is common to speak of $f \in L^1(\mathbb{R})$ being continuous, even though there is *at most* one individual function $g \in [f]$ that is continuous. And strictly speaking, we *ought not* to write $f \in L^1(\mathbb{R})$ because it is really $[f] \in L^1(\mathbb{R})$. Nevertheless, everyone writes in the simple, loose way, but bearing in mind that the notation $f \in L^1(\mathbb{R})$ is just a common abuse of notation for the sake of simplicity of expression.

EXERCISES

5.38   Let $f = 1_{\mathbb{Q} \cap [0,1]} \in L^1[0,1]$ with respect to Lebesgue measure. Prove that f is both continuous and differentiable on its domain in the sense that there is a unique member of the equivalence class of f with those properties.

5.39   Let $f = 1_{[0,1]} \in L^1(\mathbb{R})$. Prove that f is neither continuous nor differentiable, in the sense that the equivalence class of f contains no function with either of those two properties.

Definition 5.5.4   A set $S \subseteq L^1(X, \mathfrak{A}, \mu)$ is said to be *dense* in $L^1(X, \mathfrak{A}, \mu)$ provided that for each $f \in L^1(X, \mathfrak{A}, \mu)$, and for each $\epsilon > 0$, there exists $s \in S$ such that $\|s - f\|_1 < \epsilon$.

The following theorem is frequently useful for proving theorems about integrable functions.

Theorem 5.5.1   *The space $\mathfrak{S}_0$ of special simple functions is dense in the space $L^1(X, \mathfrak{A}, \mu)$.*

The reader should note how we have made the conventional abuse of language in the statement of this theorem: $\mathfrak{S}_0$ is not actually contained in $L^1(X, \mathfrak{A}, \mu)$, because the latter space is a set of equivalence classes of function—not of functions themselves. Thus it is really not the set of special simple functions but the set of equivalence classes of those functions that is dense in L^1! However, we will abuse language in this way for simplicity of expression, and this should not cause difficulty if the reader remains alert to it.

Proof: Let $f \in L^1(X, \mathfrak{A}, \mu)$ and write $f = f^+ - f^-$, as in Definition 5.2.5. We know that there exists a function $\phi^+ \in \mathfrak{S}_0$ such that $0 \leqslant \phi^+ \leqslant f^+$, and also such that

$$\int_X \phi^+ \, d\mu > \int_X f^+ \, d\mu - \frac{\epsilon}{2}.$$

We define ϕ^- similarly. Let $\phi_= \phi^+ - \phi^-$, and apply the triangle inequality as follows:

$$\|f - \phi\| \leqslant \|f^+ - \phi^+\| + \|f^- - \phi^-\| < \epsilon.$$

■

In the case of the real line with Lebesgue measure, there is a very special subset of $\mathfrak{S}_0$ called the *step functions*.

Definition 5.5.5 Define a *step function* on an interval $[a, b] \subset \mathbb{R}$ as follows. We call σ a step function if there exists a partition $a = x_0 < x_1 < \cdots < x_n = b$ of $[a, b]$ into finitely many contiguous closed intervals such that $\sigma(x) = c_i$, a constant for all $x \in (x_{i-1}, x_i)$, $i = 1, \ldots, n$. Thus σ is constant on each *open* interval (x_{i-1}, x_i). The values of σ at $x_0, x_1, \ldots x_n$ are arbitrary. Let S denote the family of all step functions on the real line.

The reader will show in Exercise 5.41 that the vector space S of all step functions is dense in $L^1(\mathbb{R})$. The latter fact is very useful for proving many theorems in real analysis.

Definition 5.5.6 In a normed linear space V, a sequence of vectors v_n is called a *Cauchy sequence* provided that for each $\epsilon > 0$, there exists an $N \in \mathbb{N}$ such that, for all m and n greater than or equal to N, we have $\|v_n - v_m\| < \epsilon$. A normed linear space is called *complete* provided that, for each Cauchy sequence v_n in V, there exists $v \in V$ such that $v_n \to v$, which means that $\|v_n - v\| \to 0$. A complete normed real linear space is called a *real Banach space*, and a complete normed complex linear space is called a *Banach space*.

EXERCISE

5.40 Give an example of a Cauchy sequence of functions

$$f_n \in L^1[0, 1]$$

such that there does not exist *any* point $x \in [0, 1]$ for which $f_n(x)$ converges. Prove that your example has the properties that are claimed.

Theorem 5.5.2 *Let $(X, \mathfrak{A}, \mu)$ be a measure space. Then $L^1(X, \mathfrak{A}, \mu)$ is a complete normed linear space.*[47]

Proof: Let f_n be a Cauchy sequence in $L^1(X, \mathfrak{A}, \mu)$. We must show there exists $f \in L^1(X, \mathfrak{A}, \mu)$ such that $\|f_n - f\|_1 \to 0$ as $n \to \infty$. The main difficulty in the proof is to find a suitable function f in $L^1(X, \mathfrak{A}, \mu)$. The reason this is challenging is that a Cauchy sequence in $L^1(X, \mathfrak{A}, \mu)$ need not converge pointwise at any point $x \in X$, as is shown in Exercise 5.40. In order to remedy this difficulty, we will prove that if a sequence of L^1-functions is sufficiently *rapidly Cauchy*, then it must converge pointwise almost everywhere.

---

[47]That is, $L^1(X, \mathfrak{A}, \mu)$ is a real Banach space. In Section 5.6 we will show how to deal with complex-valued L^1-functions, and then $L^1(X, \mathfrak{A}, \mu)$ will be a full-fledged Banach space in the complex sense.

Note that since f_n is Cauchy, there exists an increasing sequence of natural numbers $n_k \in \mathbb{N}$ such that if n and m are greater than or equal to n_k, then

$$\|f_n - f_m\|_1 < \frac{1}{4^k}. \tag{5.5}$$

In particular,

$$\|f_{n_k} - f_{n_{k+1}}\|_1 < \frac{1}{4^k} \tag{5.6}$$

for each $k \in \mathbb{N}$. The following lemma will be very helpful.

Lemma 5.5.1 (Pointwise Convergence Lemma) *Let $(X, \mathfrak{A}, \mu)$ be a measure space, and suppose a sequence of functions $g_k \in L^1(X, \mathfrak{A}, \mu)$ satisfies the equation*

$$\|g_k - g_{k+1}\|_1 < \frac{1}{4^k}$$

for each $k \in \mathbb{N}$. Then there exists a function $g \in L^1(X, \mathfrak{A}, \mu)$ such that $g_k(x) \to g(x)$ for almost all values of x. Moreover, $g_k \to g$ in the sense of L^1-norm convergence, meaning that $\|g_k - g\|_1 \to 0$.

Proof: We think of the sequence g_k as being very *rapidly* Cauchy. Note that the set

$$A_k = \left\{ x \,\middle|\, |g_k(x) - g_{k+1}(x)| \geq \frac{1}{2^k} \right\}$$

is measurable, since $|g_k - g_{k+1}|$ is $\mathfrak{A}$-measurable. Furthermore,

$$\frac{1}{2^k} \mu(A_k) \leq \int_X |g_k - g_{k+1}| \, d\mu < \frac{1}{4^k},$$

so that $\mu(A_k) < \frac{1}{2^k}$ for each $k \in \mathbb{N}$. Next, we define

$$N = \limsup A_k = \bigcap_{p=1}^{\infty} \bigcup_{k=p}^{\infty} A_k,$$

and we note that N is the set of all points x that lie in infinitely many of the sets A_k. Furthermore

$$\mu \left(\bigcup_{k=p}^{\infty} A_k \right) \leq \sum_{k=p}^{\infty} \frac{1}{2^k}$$

$$= \frac{1}{2^{p-1}} \to 0$$

as $p \to \infty$. It follows that $\mu(N) = 0$, so that N is an $(\mathfrak{A}, \mu)$-null set. We will show that the sequence $g_k(x)$ converges for each $x \in X \backslash N$.

If $x \notin N$, then there exists $p \in \mathbb{N}$ such that $k \geq p$ implies that

$$|g_k(x) - g_{k+1}(x)| < \frac{1}{2^k}.$$

Therefore, if k and l are greater than or equal to p, repeated application of the triangle inequality tells us that

$$|g_k(x) - g_l(x)| < \frac{1}{2^{p-1}},$$

so that $g_k(x)$ is a Cauchy sequence of real numbers. Thus the function given by

$$g(x) = \lim_{k \to \infty} g_k(x)$$

exists almost everywhere on X and is measurable.[48] Moreover, we find that

$$
\begin{aligned}
\|g_k - g\|_1 &= \int_X |g_k - g| \, d\mu \\
&= \int_X \liminf_l |g_k - g_l| \, d\mu \\
&\leq \liminf_l \int_X |g_k - g_l| \, d\mu \\
&\leq \frac{1}{4^{p-1}},
\end{aligned}
$$

by Fatou's theorem.[49] Thus $g_k - g \in L^1(X)$, which implies that $g \in L^1(X)$ and that $g_k \to g$ in the L^1-norm. This proves the pointwise convergence lemma. ∎

To finish the proof of Theorem 5.5.2, we substitute f_{n_k} from Equation (5.5) for g_k in the pointwise convergence lemma.

We need to show the convergence in the L^1-norm of f_n to f, playing the role of g from the lemma. Note that if $n \geq n_k$, then we have $\|f_n - f_{n_k}\|_1 < \frac{1}{4^k}$. It follows that if $n \geq n_k$, then we have

$$
\begin{aligned}
\|f_n - f\|_1 &\leq \|f_n - f_{n_k}\|_1 + \|f_{n_k} - f\|_1 \\
&< \frac{2}{4^{k-1}} \to 0
\end{aligned}
$$

as $k \to \infty$. Hence $\|f_n - f\|_1 \to 0$ as $n \to \infty$, and $L^1(X)$ is complete. ∎

---

[48] If $(X, \mathfrak{A}, \mu)$ happens to be a complete measure space, then it is clear that g is measurable. If the measure space is not complete, we can define g to be constant on the null set N of nonconvergence pointwise, and again g will be measurable. For the purposes of this proof, we need to find only one L^1-function that can serve as the limit for the original Cauchy sequence.

[49] Fatou's theorem is useful here, since we do not have a dominating function that would be required for the Lebesgue Dominated Convergence theorem.

EXERCISES

5.41   Prove the following three handy lemmas.

 a) On the real line $\mathbb{R}$, prove that the vector space S of step functions is dense in $\mathfrak{S}_0$ with respect to the L^1-norm. Explain why this implies density of S in $L^1(\mathbb{R})$ as well. (Hint: If A is a measurable set, prove that the indicator function 1_A is the limit of a sequence of step functions σ_n. That is, $\|1_A - \sigma_n\|_1 \to 0$ as $n \to \infty$. You can use the result of Exercise 3.4.)

 b) Prove that each function $f \in L^1(\mathbb{R})$ is equal almost everywhere to a Borel measurable function.

 c) Let $g : \mathbb{R} \to \mathbb{R}$ and suppose that g is equal to a Borel measurable function ϕ except on a set of Lebesgue measure zero. Prove that g is Lebesgue measurable.

5.42   Let $f \in L^1(\mathbb{R})$ and define the Fourier sine transform $\hat{f}$ by

$$\hat{f}(\alpha) = \int_{\mathbb{R}} f(x) \sin \alpha x \, dl(x)$$

for each $\alpha \in \mathbb{R}$. Prove that $\hat{f}(\alpha) \to 0$ as $\alpha \to \infty$. This statement is known as the *Riemann-Lebesgue* lemma. (Hint: Treat first the special case in which f is the indicator function of an interval.)

5.43   If $f \in L^1(\mathbb{R})$, define $f_t \in L^1(\mathbb{R})$ by

$$f_t(x) = f(x + t),$$

the *translate* of f by t. For each $t \in \mathbb{R}$, define $\phi(t)$ to be the linear transformation

$$\phi(t) : L^1(\mathbb{R}) \to L^1(\mathbb{R}),$$

given by $\phi(t) : f \to f_t$ for each $t \in \mathbb{R}$ and for each $f \in L^1(\mathbb{R})$.

 a) Prove that ϕ is a *homomorphism* from the additive group of real numbers into the group $\mathcal{L}\left(L^1(\mathbb{R})\right)$ of linear transformations of the vector space $L^1(\mathbb{R})$ into itself. The group $\mathcal{L}\left(L^1(\mathbb{R})\right)$ is equipped with the operation of composition.[50]

 b) Fix $t \in \mathbb{R}$ and show that

$$\|\phi(t)f\|_1 = \|f\|_1$$

for each $f \in L^1(\mathbb{R})$. (For those who know what is meant by the norm of a linear transformation, this says that $\phi(t)$ is a bounded—hence continuous—linear self-transformation of $L^1(\mathbb{R})$.[51])

---

[50]This part is a purely algebraic question concerning the group operations.
[51]This part is only for those students who know how to put a norm on the vector space of continuous linear transformations of a normed linear space. See, for example, [20] and [21].

c) Prove that $\phi(t)f$ is continuous at the point $t = 0$. This means that for each fixed f in $L^1(\mathbb{R})$, we have

$$\|\phi(t)f - \phi(0)f\|_1 \to 0$$

as $t \to 0$. Prove also continuity at each value of $t \in \mathbb{R}$. In words, this exercise says that the mapping $t \to \phi(t)f$ is a continuous mapping from $\mathbb{R} \to L^1(\mathbb{R})$.

d) Fix any $t > 0$, no matter how small. Show that there exists a function f in $L^1(\mathbb{R})$ such that $\|f\|_1 = 1$ yet

$$\|\phi(t)f - f\|_1 = 2.$$

Thus $\|\phi(t) - \phi(0)\|$ *fails* to converge to 0 as $t \to 0$, using the concept of the norm of a linear transformation of a normed linear space. [52]

5.44 Denote $[0,1] \cap \mathbb{Q} = \{q_k \mid k \in \mathbb{N}\}$, the countable set of all rational numbers in $[0,1]$. Let $f_k : [0,1] \to \mathbb{R}^*$ be defined almost everywhere by

$$f_k(x) = \frac{1}{\sqrt{|x - q_k|}}$$

for each $k \in \mathbb{N}$. Note that the graph of f_k has a vertical asymptote at $x = q_k$. (One may decide to define $f(q_k)$ arbitrarily, but this is not relevant to the problem.)

a) Show that $f_k \in L^1[0,1]$ and find an upper bound for $\|f_k\|_1$ that is independent of k.

b) Let

$$S_n = \sum_{k=1}^{n} \frac{f_k}{2^k}$$

for each $n \in \mathbb{N}$. Prove that S_n is a Cauchy sequence in the L^1-norm, and thus that $S = \lim_{n \to \infty} S_n \in L^1[0,1]$.

c) Prove for each $x \in [0,1]$ that $S_n(x)$ either diverges to infinity or else converges, and that $\lim_{n \to \infty} S_n(x) < \infty$ for almost all x.

d) Show that S_1 is *improperly* Riemann integrable in the sense of elementary calculus. Show that S_n may be considered improperly Riemann integrable in a plausible sense that should be explained. Show that $\lim_{x \to q} S(x)$ is infinite at each rational point $q \in E$. Can you find any reasonable way to describe S as improperly Riemann integrable? [53]

---

[52] The point here is that continuity in analysis is a very delicate issue indeed. It is very much a matter of what is the mapping, what is the domain (and its norm or topology), and what is the range (and its norm or topology).

[53] This example can be interpreted also as an instance of Lebesgue Dominated Convergence of the sequence of partial sums, regarding S itself as the dominating function. Unlike many examples used for Lebesgue Dominated Convergence, in this one the dominating function is *not* even *improperly* Riemann integrable. This exercise shows that Lebesgue integration obviates the need for a concept of improper integration and goes very much farther than the concept of improper integration used for the Riemann integral.

5.45   Suppose $f_n \in L^1(X, \mathfrak{A}, \mu)$ and $\|f_n\|_1 \to 0$ as $n \to \infty$. Prove that $f_n \to 0$ in measure as well.

5.6   COMPLEX-VALUED FUNCTIONS

What we have learned thus far about $L^1(X, \mathfrak{A}, \mu)$ can be extended easily to complex-valued functions. (This should not be confused with complex analysis, in which the independent variable x is replaced by a complex independent variable z.) In contexts in which it would not be obvious in what set the values of a function lie, it is common to write such symbols as $L^1(X, \mathbb{R})$ or $L^1(X, \mathbb{C})$ to indicate (with the last entry) what type of value the function has. In more advanced subjects, such as harmonic analysis on Lie groups and representation theory, one learns about $L^1(X, \mathcal{H})$, in which the functions take their values in a complex Hilbert space $\mathcal{H}$ rather than in the real or complex field.

If $f : X \to \mathbb{C}$, we write $f(x) = u(x) + iv(x)$, where the *real part* of $f(x)$ is denoted by

$$u(x) = \Re f(x)$$

and the *imaginary part* is denoted by

$$v(x) = \Im f(x),$$

both of which are real-valued.   Note that the complex modulus

$$|f(x)| = \sqrt{(\Re f(x))^2 + (\Im f(x))^2}.$$

Definition 5.6.1 A complex-valued function $f : (X, \mathfrak{A}, \mu) \to \mathbb{C}$ is called *measurable* provided that both $\Re f$ and $\Im f$ are measurable. We say that $f \in L^1(X, \mathbb{C})$ provided that both $\Re f \in L^1(X, \mathbb{R})$ and $\Im f \in L^1(X, \mathbb{R})$. When the latter conditions are satisfied, we write that

$$\int_X f \, d\mu = \int_X \Re f \, d\mu + i \int_X \Im f \, d\mu$$

and we define the L^1-norm of f to be

$$\|f\|_1 = \int_X |f| \, d\mu = \int_X \sqrt{(\Re f)^2 + (\Im f)^2} \, d\mu.$$

EXERCISES

5.46   Let $(X, \mathfrak{A}, \mu)$ be a measure space.

     a) Prove that $f \in L^1(X, \mathbb{C})$ if and only if both $\Re f$ and $\Im f$ are in $L^1(X, \mathbb{R})$.

     b) Prove that $f \in L^1(X, \mathbb{C})$ if and only if $\|f\|_1 < \infty$, where $\|f\|_1 = \int_X |f| \, d\mu$. (Here $|f|$ denotes the *modulus* of f.)

c) Prove that if $f \in L^1(X, \mathbb{C})$, then

$$\left| \int_X f \, d\mu \right| \leq \int_X |f| \, d\mu.$$

(Hint: Denote $\int_X f \, d\mu = Re^{i\Theta}$ where $R = \left| \int_X f \, d\mu \right| \geq 0$ and $\Theta \in \mathbb{R}$.)

d) Prove that $L^1(X, \mathbb{C})$ is a complete normed linear space.

5.47   The Fourier transform of a complex-valued function $f \in L^1(\mathbb{R}, \mathbb{C})$ with respect to Lebesgue measure is given by

$$\widehat{f}(\alpha) = \int_{\mathbb{R}} f(x) \, e^{-2\pi i \alpha x} \, dx$$

for each $\alpha \in \mathbb{R}$.

a) Prove that $\left| \widehat{f}(\alpha) \right| \leq \|f\|_1$ for all $\alpha \in \mathbb{R}$.

b) Prove that $\widehat{f} \in C(\mathbb{R})$, the space of continuous functions on $\mathbb{R}$.

c) Prove the Riemann-Lebesgue lemma: $\widehat{f}(\alpha) \to 0$ as $|\alpha| \to \infty$. (Hint: See Exercise 5.42.)

5.48   Denote by l_1 the vector space of all complex sequences z that are *absolutely summable*:

$$\sum_{j \in \mathbb{N}} |z_j| < \infty. \tag{5.7}$$

Call the sum in Equation (5.7) the l_1-norm of the sequence z, denoted by $\|a_j\|_1$. Prove that l_1 with the given norm is a complete normed vector space.

CHAPTER 6

PRODUCT MEASURES AND FUBINI'S THEOREM

All students of mathematics learn in elementary calculus to evaluate a double integral by iteration. The theorem justifying this process is called *Fubini's theorem*. However, the cleanest and simplest form of Fubini's theorem appears for the first time with the Lebesgue integral. We will see in this chapter that Fubini's theorem is instrumental in proving many other important theorems of real analysis.

6.1 PRODUCT MEASURES

We assume here that we are given two measure spaces, $(X, \mathfrak{A}, \lambda)$ and $(Y, \mathfrak{B}, \mu)$. The two measure spaces are permitted to be identical. We intend to construct the product measure on a suitable σ-field contained in the power set of the Cartesian product $Z = X \times Y$. By a *rectangular set* R in Z we mean any set of the form $R = A \times B$, where $A \in \mathfrak{A}$ and $B \in \mathfrak{B}$. We will take as the family of *elementary sets* for the product measure

$$\mathfrak{E} = \left\{ E = \dot{\bigcup}_{i=1}^{n} R_i \,\middle|\, R_i = A_i \times B_i, \, A_i \in \mathfrak{A}, \, B_i \in \mathfrak{B} \right\}, \tag{6.1}$$

where the rectangles R_i are *mutually disjoint* and n is an arbitrary natural number.

Measure and Integration: A Concise Introduction to Real Analysis. By Leonard F. Richardson
Copyright © 2009 John Wiley & Sons, Inc.

EXERCISE

6.1   Show that the set $\mathfrak{E}$ of Equation (6.1) is a field of subsets of $Z = X \times Y$. (Be sure to check closure under complementation.)

Definition 6.1.1   Define the product measure

$$\nu(E) = \sum_{i=1}^{n} \lambda(A_i)\mu(B_i)$$

for each elementary set $E \in \mathfrak{E}$ as defined by Equation (6.1).

This definition requires justification, because the decomposition given in Equation (6.1) is not unique. Suppose

$$E = \overset{m}{\underset{i=1}{\overset{\cdot}{\bigcup}}} A_i \times B_i = \overset{n}{\underset{j=1}{\overset{\cdot}{\bigcup}}} C_j \times D_j.$$

It follows from the finite additivity of each of the measures λ and μ that no rectangular set of *infinite* measure can be expressed as a union of finitely many rectangular sets of finite measure. Thus

$$\sum_{i=1}^{m} \lambda(A_i)\mu(B_i) = \infty \Leftrightarrow \sum_{j=1}^{n} \lambda(C_j)\mu(D_j) = \infty.$$

We can use the integral of a special simple function to rephrase Definition 6.1.1 in a way that expresses concisely why that definition is independent of the decomposition in the case in which $\sum_{i=1}^{m} \lambda(A_i)\mu(B_i) < \infty$.

Definition 6.1.2   If $S \subseteq X \times Y$, a Cartesian product space, we define the x-section of S by

$$_xS = \{y \mid (x,y) \in S,\ y \in Y\}$$

and the y-section by

$$S_y = \{x \mid (x,y) \in S,\ x \in X\}.$$

If $E \in \mathfrak{E}$, the field of elementary sets, then

$$E = \overset{n}{\underset{i=1}{\overset{\cdot}{\bigcup}}} R_i = \overset{n}{\underset{i=1}{\overset{\cdot}{\bigcup}}} A_i \times B_i.$$

Define the *x-section function* by

$$f_E(x) = \mu(_xE) = \sum_{\{i \mid x \in A_i\}} \mu(B_i) \tag{6.2}$$

$$= \sum_{i=1}^{n} 1_{A_i}(x)\,\mu(B_i).$$

If $\displaystyle\sum_{i=1}^{m} \lambda(A_i)\mu(B_i) < \infty$, then we see from Equations (6.2) that f_E is a nonnegative special simple function on X. Moreover,

$$\int_X f_E \, d\lambda = \sum_{i=1}^{n} \lambda(A_i)\,\mu(B_i) = \nu(E). \tag{6.3}$$

From Equation (6.3), it is clear that Definition 6.1.1 is independent of the decomposition of the elementary set into a disjoint union of rectangular sets. In fact, if we consider two such decompositions, then we will have two different expansions of the same simple cross-section function f_E, so the result of the integral $\int_X f_E \, d\lambda$ must be the same either way.

Furthermore, if E and E' are any two *mutually disjoint* elementary sets in $X \times Y$, then the x-section function

$$f_{E \cup E'} = f_E + f_{E'}.$$

This establishes that

$$\nu\left(E \dot\cup E'\right) = \nu(E) + \nu\left(E'\right),$$

since integration over X is a linear function of the integrand.

EXERCISE

6.2   Let $E \in \mathfrak{E}$, the field of elementary sets in the product of two measure spaces, $(X, \mathfrak{A}, \lambda)$ and $(Y, \mathfrak{B}, \mu)$. Prove that the product measure $\nu = \lambda \times \mu$ is given on $\mathfrak{E}$ by

$$\nu(E) = \int_Y \lambda\left(E_y\right) d\mu.$$

Our next goal is to show that, under suitable hypotheses, ν can be extended to a countably additive measure on $\mathfrak{C} = \mathbb{B}(\mathfrak{E})$, the σ-field generated by $\mathfrak{E}$.

Theorem 6.1.1 *Let $(X, \mathfrak{A}, \lambda)$ and $(Y, \mathfrak{B}, \mu)$ be any two measure spaces. There exists a countably additive measure ν defined on $\mathfrak{C} = \mathbb{B}(\mathfrak{E})$ such that*

$$\nu(A \times B) = \lambda(A)\mu(B)$$

for all $A \in \mathfrak{A}$ and $B \in \mathfrak{B}$. Moreover, ν is unique, provided that λ and μ are both σ-finite.

Proof: Uniqueness will follow at the end from Exercise 2.21, since ν will be σ-finite if both λ and μ are σ-finite. Existence depends upon proving countable additivity *within* $\mathfrak{E}$.

Each elementary set is a *disjoint* union of finitely many rectangular sets of the form

$$E = \dot{\bigcup}_{i=1}^{n} A_i \times B_i,$$

with each A_i and each B_i measurable. To prove the countable additivity within $\mathfrak{E}$, it will suffice to prove for each rectangular set

$$R = A \times B = \dot{\bigcup}_{n \in \mathbb{N}} A_n \times B_n$$

that

$$\nu(A \times B) = \sum_{n \in \mathbb{N}} \lambda(A_n)\mu(B_n).$$

To this end, we define

$$f_n(x) = \mu\left({}_x\left(\bigcup_{i=1}^{n}(A_i \times B_i)\right)\right)$$

so that f_n is a nonnegative simple function. Note that although the sets B_i need not be disjoint, we do have

$$_x(A \times B) = \dot{\bigcup}_{n \in \mathbb{N}} {}_x(A_n \times B_n),$$

which is a disjoint union. Since μ is countably additive, it follows that the sequence f_n increases monotonically toward the limit f_R, where

$$f_R(x) = \mu\left({}_xR\right) = \begin{cases} \mu(B) & \text{if } x \in A, \\ 0 & \text{if } x \notin A. \end{cases}$$

By the Monotone Convergence theorem (5.4.1) we see that

$$\begin{aligned}
\nu(R) = \int_A f \, d\lambda &= \lim_{n \to \infty} \int_A f_n \, d\lambda \\
&= \lim_{n \to \infty} \sum_{i=1}^{n} \lambda(A_i)\,\mu(B_i) \\
&= \sum_{n \in \mathbb{N}} \lambda(A_i)\,\mu(B_i),
\end{aligned}$$

and this is true even if $\nu(R) = \infty$. ∎

Note that in the preceding proof we have applied the Monotone Convergence theorem only to nonnegative measurable functions defined on X, on which there is already a countably additive measure λ. (We are not applying Monotone Convergence to functions defined on the product space, for which we are in the process of establishing the existence of a countably additive measure.) We remark that we did not need to use the hypothesis of completeness that appeared in the Monotone Convergence theorem, because in the preceding proof, f_n converges *everywhere* to f—not merely almost everywhere.

Definition 6.1.3 Let $(X, \mathfrak{A}, \lambda)$ and $(Y, \mathfrak{B}, \mu)$ be any two measure spaces. We will denote the Cartesian product

$$\mathfrak{A} \times \mathfrak{B} = \{A \times B \mid A \in \mathfrak{A}, B \in \mathfrak{B}\},$$

and the field that it generates is called $\mathfrak{E}$. But we will denote by

$$\mathfrak{A} \otimes \mathfrak{B} = \mathbb{B}(\mathfrak{A} \times \mathfrak{B})$$

the Borel field generated by the field $\mathfrak{E}$. Finally, we will denote by

$$\mathfrak{A} \bigotimes \mathfrak{B}$$

the *completion* of $\mathfrak{A} \otimes \mathfrak{B}$ with respect to the product measure $\lambda \times \mu$.

It is reasonable to wonder whether it is necessary to form the completion $\mathfrak{A} \bigotimes \mathfrak{B}$ if both $\mathfrak{A}$ and $\mathfrak{B}$ happen to be complete families of measurable sets for their respective measures. The answer—that the completion needs to be formed—is confirmed by the following special case.

Let $X = Y = \mathbb{R}$ and $Z = X \times Y = \mathbb{R}^2$, the Euclidean plane. Here we will assume that $\lambda = \mu = l$, Lebesgue measure on the line. And $\mathfrak{A} = \mathfrak{B} = \mathfrak{L}(\mathbb{R})$ will be the σ-field of all Lebesgue measurable sets in the line. Recall that we have constructed earlier the family $\mathfrak{B}(\mathbb{R}^n)$ of Borel sets in $\mathbb{R}^n$ and the family $\mathfrak{L}(\mathbb{R}^n)$ of Lebesgue measurable sets in $\mathbb{R}^n$ for each $n \in \mathbb{N}$.

Theorem 6.1.2 *With the notations of Definition 6.1.3, we have*

$$\mathfrak{B}\left(\mathbb{R}^2\right) \overset{(i)}{\subsetneqq} \mathfrak{L}(\mathbb{R}) \otimes \mathfrak{L}(\mathbb{R}) \overset{(ii)}{\subsetneqq} \mathfrak{L}(\mathbb{R}) \bigotimes \mathfrak{L}(\mathbb{R}) \overset{(iii)}{=} \mathfrak{L}\left(\mathbb{R}^2\right).$$

Moreover, if we denote by l_2 the Lebesgue measure defined on $\mathbb{R}^2$ as in Definition 3.5.1, and by l Lebesgue measure on $\mathbb{R}$, then $l \times l = l_2$.

Proof: We will justify the claims numbered (i), (ii), and (iii) in order.

i. Each Cartesian product of two *intervals* lies in $\mathfrak{L}(\mathbb{R}) \times \mathfrak{L}(\mathbb{R})$. Thus the σ-field $\mathfrak{B}(\mathbb{R}^2)$ that they generate is contained in $\mathfrak{L}(\mathbb{R}) \otimes \mathfrak{L}(\mathbb{R})$. Observe that the family

$$S = \left\{ S \subseteq \mathbb{R}^2 \,\middle|\, S_y \in \mathfrak{B}(\mathbb{R}) \,\forall y \in \mathbb{R} \right\}$$

is a σ-field in the plane containing the elementary sets of the plane, as defined in Section 3.5. Thus it contains all the Borel sets of the plane. On the other hand, by Remark 3.3.1, there exists a set

$$E \in \mathfrak{L}(\mathbb{R}) \backslash \mathfrak{B}(\mathbb{R}),$$

so that

$$E \times \mathbb{R} \in \mathfrak{L}(\mathbb{R}) \otimes \mathfrak{L}(\mathbb{R}).$$

Hence there is a set $E \times \mathbb{R}$ that is in $\mathfrak{L}(\mathbb{R}) \otimes \mathfrak{L}(\mathbb{R})$ but not in $\mathfrak{B}\left(\mathbb{R}^2\right)$. This proves the first (improper) containment.[54]

ii. For the second improper containment, let M be any nonmeasurable subset of $\mathbb{R}$. Then the measure $(l \times l)(\{x\} \times M) = 0$ for each singleton set $\{x\}$ in $\mathbb{R}$, being a subset of a null set. Hence

$$\{x\} \times M \in \mathfrak{L}(\mathbb{R}) \bigotimes \mathfrak{L}(\mathbb{R}).$$

Thus it would suffice to show that $\{x\} \times M \notin \mathfrak{L}(\mathbb{R}) \otimes \mathfrak{L}(\mathbb{R})$. Consider the class $\mathcal{S}$ of all sets $E \subset \mathbb{R}^2$ for which $_xE \in \mathfrak{L}(\mathbb{R})$ for a fixed $x \in \mathbb{R}$. Then it is easy to check that $\mathcal{S}$ is a monotone class, which implies that $\mathcal{S}$ contains the σ-field $\mathfrak{L}(\mathbb{R}) \otimes \mathfrak{L}(\mathbb{R})$. But since the set $\{x\} \times M$ lacks this property, it follows that $\{x\} \times M \notin \mathfrak{L}(\mathbb{R}) \otimes \mathfrak{L}(\mathbb{R})$.

iii. The space $\mathfrak{L}\left(\mathbb{R}^2\right)$ contains both $\mathfrak{L}(\mathbb{R}) \otimes \mathfrak{L}(\mathbb{R})$[55] and $\mathfrak{B}(\mathbb{R}^2)$. Since $\mathfrak{L}\left(\mathbb{R}^2\right)$ is the (minimal) completion of $\mathfrak{B}\left(\mathbb{R}^2\right)$, it must be also the unique minimal completion of $\mathfrak{L}(\mathbb{R}) \otimes \mathfrak{L}(\mathbb{R})$, as is $\mathfrak{L}(\mathbb{R}) \bigotimes \mathfrak{L}(\mathbb{R})$.

Finally, it is clear that $l \times l(R) = l_2(R)$ for each rectangle

$$R = [a, b] \times [c, d].$$

Thus the two measures agree as well on the Borel sets and on the Lebesgue measurable sets in the plane $\mathbb{R}^2$. As we have seen above, the family of Lebesgue measurable sets on $\mathbb{R}^2$ coincides with the family of Lebesgue measurable sets in $\mathbb{R} \times \mathbb{R}$. ∎

6.2 FUBINI'S THEOREM

Theorem 6.2.1 (Fubini's Theorem—First Form) *Let* $(X, \mathfrak{A}, \mu)$ *and* $(Y, \mathfrak{B}, \nu)$ *be complete* σ-*finite measure spaces. Let* $\mathfrak{C} = \mathfrak{A} \bigotimes \mathfrak{B}$. *Then for each* $\mu \times \nu$-*measurable set* $C \in \mathfrak{C}$, *the section* $_xC$ *is measurable for almost all* x, *the function* $f_C(x) = \nu(_xC)$ *is* $\mathfrak{A}$-*measurable, and*

$$(\mu \times \nu)(C) = \int_X f_C(x) \, d\mu(x). \tag{6.4}$$

Proof: Note that $(X \times Y, \mathfrak{A} \bigotimes \mathfrak{B}, \mu \times \nu)$ must be σ-finite as well, so that

$$X \times Y = \bigcup_{n \in \mathbb{N}} X_n \times Y_n,$$

---

[54]The strict containment (i) could be proven in another way, by using a cardinality argument. The transfinite cardinal number of the right-hand side is greater than that of the left.

[55]It is not hard to see that because $\mathfrak{L}\left(\mathbb{R}^2\right)$ contains the products of intervals, and because it is a σ-algebra, it must contain also all products of the form $B \times \mathbb{R}$ or $\mathbb{R} \times B$, where B is a Borel set in the line. Thus it contains also the products of Borel sets. However, each Lebesgue measurable set is sandwiched between two Borel sets of the same measure. This, together with the completeness of $\mathfrak{L}(\mathbb{R}^2)$, justifies the claim.

where $\mu(X_n)\nu(Y_n) < \infty$ and $X_n \times Y_n$ is an ascending chain of sets that are rectangular and thus elementary and Borel as well. We remark that the first form of Fubini's theorem expresses the product measure $\mu \times \nu$ of a set $C \in \mathfrak{C}$ as the integral with respect to μ of the ν-measures of the x-sections of C. This includes the possibility of both sides of Equation (6.4) being infinite.

Observe that the theorem follows easily from the definition of the product measure on the field $\mathfrak{E}$ of elementary sets in the special case in which C is an elementary set in the product space. For the latter sets, each section $_xC$ is measurable as well, being a union of finitely many measurable subsets of Y.

Next, we wish to prove the theorem for the case in which C is a Borel set, meaning that $C \in \mathfrak{A} \otimes \mathfrak{B}$. By Theorem 2.1.2, we see that Fubini's theorem would be true for all Borel sets C provided that the family $\mathcal{F}$ of sets for which the theorem is true is a *monotone class.*

1. We will prove first that $\mathcal{F}$ is closed under the operation of forming the union of an increasing chain of sets in $C_n \in \mathcal{F}$. This conclusion will follow from the Monotone Convergence theorem, together with the fact that the pointwise limit almost everywhere of measurable functions must be measurable in a complete measure space. Let $C = \bigcup_n C_n$, the union of an ascending chain of sets in $\mathcal{F}$. We let $f_n(x) = \nu(_xC_n)$, which is a monotone increasing sequence of measurable functions defined almost everywhere. We observe that $f_n \to f_C$, where $f_C(x) = \nu(_xC)$, because $_xC$ is the union of the increasing chain $_xC_n$ of measurable sets and because ν is countably additive. Hence

$$\int_X f \, d\mu = \lim_{n \to \infty} \int_X f_n \, d\mu$$
$$= \lim_{n \to \infty} (\mu \times \nu)(C_n)$$
$$= (\mu \times \nu)(C),$$

with the final equality following from the countable additivity of the product measure $\mu \times \nu$.

2. For a decreasing nest C_n, we limit ourselves first to a typical subspace,

$$S_N = X_N \times Y_N,$$

of finite product measure. Thus we assume at first that each $C_n \subseteq S_N$, which has finite measure. Define $f_n(x) = \nu(_xC_n)$ as before, and observe that f_1 is an integrable function[56] dominating the decreasing sequence f_n, and $f_n(x) \to f_C(x) = \nu(_xC)$ for almost all x. Then the Lebesgue Dominated

---

[56]The function f_n is integrable because it is a bounded, measurable function on a space X_N of finite measure. Note that f_n is bounded because $\nu(Y_N) < \infty$.

Convergence theorem implies that

$$
\int_X f_C \, d\mu = \lim_{n \to \infty} \int_X f_n \, d\mu
$$
$$
= \lim_{n \to \infty} (\mu \times \nu)(C_n)
$$
$$
= (\mu \times \nu)(C).
$$

This shows that for each $N \in \mathbb{N}$ we have $\mathcal{F} \cap \mathfrak{P}(S_N)$ is a monotone class within the power set of S_N, and thus also a σ-field. Hence $\mathcal{F}$ contains all the Borel sets in S_N. However, the Borel sets of $S = X \times Y$ will be the unions of their own intersections with each of the Borel sets S_N. Hence $\mathcal{F} \supseteq \mathfrak{A} \otimes \mathfrak{B}$ by part (i), in which we showed that $\mathcal{F}$ is closed under forming unions of ascending chains. Moreover, every section $_xC$ of a Borel set must be measurable, as shown in the proof of part (ii) of Theorem 6.1.2.

To complete the proof of Theorem 6.2.1 for all measurable sets $C \in \mathfrak{A} \otimes \mathfrak{B}$, recall that each measurable set differs from a Borel set by a null set. Thus it would suffice to prove that Fubini's theorem applies to all C that are null sets with respect to the measure $\mu \times \nu$. If $\mu \times \nu(C) = 0$, then there exists a Borel set $D \supseteq C$ such that $\mu \times \nu(D) = 0$ also. By the previous part of this proof, it follows that

$$
(\mu \times \nu)(D) = \int_X \nu(_xD) \, d\mu = 0.
$$

Hence $\nu(_xD) = 0$ for μ-almost all x, and thus $_xC \subseteq \, _xD$ is both a measurable set and a ν-null set for μ-almost all x.[57] It follows that $(\mu \times \nu)(C) = \int_X \nu(_xC) \, d\mu$, and the proof is complete.   ∎

Theorem 6.2.2 (Fubini's Theorem—Main Form) *Let $(X, \mathfrak{A}, \mu)$ and $(Y, \mathfrak{B}, \nu)$ be two complete σ-finite measure spaces. Suppose that*

$$
f \in L^1 \left(X \times Y, \mathfrak{A} \otimes \mathfrak{B}, \mu \times \nu \right).
$$

Then

  i. *For almost all $x \in X$, the function $f(x, \cdot) : y \to f(x, y)$ is integrable on Y.*

  ii. *For almost all $y \in Y$, the function $f(\cdot, y) : x \to f(x, y)$ is integrable on X.*

  iii. *The function $\int_Y f(x, \cdot) \, d\nu$ is integrable on X.*

  iv. *The function $\int_X f(\cdot, y) \, d\mu$ is integrable on Y.*

---

[57]One should note here that it is not necessary for each cross section of a null set in the product measure to be measurable. For example, if M is nonmeasurable in Y and if N is a null set in X, the $N \times M$ is a null set in $X \times Y$. Recall that every set of positive measure contains a nonmeasurable set by Theorem 3.4.4.

v. *We have the following equality of double integrals, with finite values:*

$$\int_X \left(\int_Y f(x, \cdot) \, d\nu \right) d\mu = \int_{X \times Y} f \, d(\mu \times \nu) \tag{6.5}$$

$$= \int_Y \left(\int_X f(\cdot, y) \, d\mu \right) d\nu.$$

Proof: Because the roles of the two variables are symmetrical, it will suffice to prove (i), (iii), and the first equality in (v). If the conclusions are true for two functions, then they are true also for the difference of the two functions. Hence it suffices to prove the statements listed for nonnegative functions, because we can write $f = f^+ - f^-$. It follows easily from Theorem 6.2.1 that the claims are true if f is the indicator function of a measurable set of finite measure. Taking finite linear combinations of such indicator functions, we see that the theorem is true if f is a special simple function. By Exercise 5.31, we know that if f is measurable and nonnegative, then f is the pointwise limit of a monotone increasing sequence of special simple functions. Thus

$$f = \lim_n \phi_n. \tag{6.6}$$

The function $f(x, \cdot)$ is a measurable nonnegative function of y for almost all x, being the pointwise limit of a sequence of functions $\phi_n(x, \cdot)$ that are measurable and integrable for almost all x. There is a different null set S_n, for each n, of values of x for which $\phi_n(x, \cdot)$ is not measurable, but the union $\bigcup_{n \in \mathbb{N}} S_n$ of countably many null sets is a null set. It follows that

$$\int_Y f(x, \cdot) \, d\nu = \lim_n \int_Y \phi_n(x, \cdot) \, d\nu$$

by the Monotone Convergence theorem for almost all x.

Thus the integral is a measurable function of x, and it follows again from monotone convergence that

$$\int_X \left(\int_Y f(x, \cdot) \, d\nu \right) d\mu = \lim_n \int_X \left(\int_Y \phi_n(x, \cdot) \, d\nu \right) d\mu$$

$$= \lim_n \int_{X \times Y} \phi_n \, d(\mu \times \nu)$$

$$= \int_{X \times Y} f \, d(\mu \times \nu).$$

Since we know now that

$$\int_X \left(\int_Y f(x, \cdot) \, d\nu \right) d\mu < \infty,$$

it follows that the inner integral of the iteration must be finite almost everywhere.

Finally, Fubini's theorem follows from the fact that f^+ and f^- are both dominated by the integrable function $|f|$ on $X \times Y$, which enables us to subtract one integral from the other since they are both finite.   ∎

The following corollary to the Fubini theorem is known as *Tonelli's theorem*. It is a useful variation of the Fubini theorem, in which the function f is given as *nonnegative*, but only *measurable* on $(X, \mathfrak{A}, \mu)$—*not* necessarily integrable. The conclusions under this modified hypothesis look similar but they are subtly altered.

Corollary 6.2.1 (Tonelli's Theorem) *Let* $(X, \mathfrak{A}, \mu)$ *and* $(Y, \mathfrak{B}, \nu)$ *be two complete σ-finite measure spaces. Suppose that* f *is a* nonnegative measurable *function on the product space* $(X \times Y, \mathfrak{A} \otimes \mathfrak{B}, \mu \times \nu)$. *Then*

 i. *For almost all* $x \in X$, *the function* $f(x, \cdot) : y \to f(x, y)$ *is measurable on* Y.

 ii. *For almost all* $y \in Y$, *the function* $f(\cdot, y) : x \to f(x, y)$ *is measurable on* X.

 iii. *The function* $\int_Y f(x, \cdot)\, d\nu$ *is measurable on* X.

 iv. *The function* $\int_X f(\cdot, y)\, d\mu$ *is measurable on* Y.

 v. *Whether the integrals are finite or infinite, we have*

$$\int_X \left(\int_Y f(x, \cdot)\, d\nu \right) d\mu = \int_{X \times Y} f\, d(\mu \times \nu) \tag{6.7}$$
$$= \int_Y \left(\int_X f(\cdot, y)\, d\mu \right) d\nu.$$

Remark 6.2.1 The reader should note that in Tonelli's theorem we do not claim that the inner integrals of the iterations (in either order) are finite almost everywhere. That conclusion would be valid provided that f is integrable on the product space, as stated in Fubini's theorem.

Proof: These hypotheses are sufficient for the validity of Equation (6.6). The remainder of the proof of Fubini's theorem is based on the Monotone Convergence theorem, which does not require integrability.   ∎

Remark 6.2.2 Tonelli's theorem has an important practical consequence. In order to use the main form of Fubini's theorem, we need a way to confirm whether or not the measurable function f is integrable. Since $|f|$ is nonnegative, we can calculate whether or not

$$\int_{X \times Y} |f|\, d(\mu \times \nu) < \infty$$

by calculating the iterated integral in either order, according to convenience. Thus, if either

$$\int_X \left(\int_Y |f(x, \cdot)|\, d\nu \right) d\mu < \infty$$

or

$$\int_Y \left(\int_X |f(\cdot, y)|\, d\mu \right) d\nu < \infty,$$

then f is an integrable function on the product space, and the full strength of the Fubini theorem can be applied to f. If one of the two orders of iteration yields a finite result, this must be true of the other order and of the integral over the product space because of Corollary 6.2.1.

Fubini's theorem is one of the most powerful tools in real analysis. The reason is that the interchange of order of iteration of a double integral is an interchange of order of two limit operations of the most delicate kind—namely, Lebesgue integration. Among the most important uses of Fubini's theorem is to prove that the inner integral of one of the two iterated integrals exists and is finite almost everywhere by establishing the finiteness of the other order of iteration when the integrand f is replaced by $|f|$. See especially Exercise 6.7.b. Several important applications are contained in the following exercises.

EXERCISES

6.3 Let $a_{ij} \in \mathbb{R}$ for all i and j in $\mathbb{N}$. Suppose that at least one of the following three sums is finite:

$$\sum_{i \in \mathbb{N}} \left(\sum_{j \in \mathbb{N}} |a_{ij}| \right), \quad \sum_{j \in \mathbb{N}} \left(\sum_{i \in \mathbb{N}} |a_{ij}| \right), \quad \text{or} \quad \sum_{(i,j) \in \mathbb{N} \times \mathbb{N}} |a_{ij}|.$$

Use Fubini's theorem to prove that

$$\sum_{i \in \mathbb{N}} \left(\sum_{j \in \mathbb{N}} a_{ij} \right) = \sum_{j \in \mathbb{N}} \left(\sum_{i \in \mathbb{N}} a_{ij} \right) = \sum_{(i,j) \in \mathbb{N} \times \mathbb{N}} a_{ij},$$

with all three sums being finite.

6.4 Suppose that f lies in $L^1(\mathbb{R}^2)$.

a) Prove that $F_n(x) = \int_0^1 f(x, y + n)\, dl(y)$ exists for almost all values of $x \in \mathbb{R}$.

b) Prove that $F_n \in L^1(\mathbb{R})$. Determine whether or not the sequence F_n has a limit in $L^1(\mathbb{R})$.

6.5 Suppose $p : \mathbb{R}^n \to \mathbb{R}$ is a *polynomial* in n real variables. Prove that the set $p^{-1}(0)$ is a Lebesgue null set in $\mathbb{R}^n$, unless p is the identically zero polynomial. (Hint: If $n = 1$ this is simple. For the inductive step, use Fubini's theorem.)

6.6

a) Suppose $h : \mathbb{R}^n \times \mathbb{R}^n \to \mathbb{R}^n$ is a measurable function such that $h^{-1}(N)$ is a Lebesgue null set for each null set N. If $f : \mathbb{R}^n \to \mathbb{R}$ is measurable, prove that $f \circ h$ is measurable.[58]

---

[58]Compare this exercise with Exercise 7.14.

b) Show that if $k(x,y) = x - y$, mapping $\mathbb{R}^n \times \mathbb{R}^n \to \mathbb{R}^n$, then k^{-1} maps null sets to null sets. (Hint: It is not automatic that k^{-1} of a null set is measurable. Prove first that k^{-1} of a Borel null set is both measurable and null. Use Theorem 6.2.1. Then show that k^{-1} maps null sets to measurable null sets.)

c) Show that if $f : \mathbb{R}^n \to \mathbb{R}$ is Lebesgue measurable function, then

$$H(x,y) = f(x - y)$$

is Lebesgue measurable on $\mathbb{R}^n \times \mathbb{R}^n$.

6.7   Suppose both f and g are L^1 functions on $\mathbb{R}^n$. In the following problems, you may use the translation invariance of both Lebesgue measure (Exercises 3.6 and 3.24) and the Lebesgue integral on $\mathbb{R}^n$ (Exercise 5.7), as well as Exercises 6.6 and 5.41.

a) Show that

$$h(x, y) = f(x - y)g(y)$$

is an L^1 function on $\mathbb{R}^{2n}$. Be sure to explain why h is measurable. You can use the result of Exercise 6.6.[59]

b) Show that the *convolution*, denoted and defined by

$$f * g(x) = \int_{\mathbb{R}^n} f(x - y)g(y)\, dl(y),$$

is defined almost everywhere in x.

c) Show that $f * g$ is an integrable function on $\mathbb{R}^n$.

d) Show that $\|f * g\|_1 \leq \|f\|_1 \|g\|_1$.

e) Show that $f * g = g * f$. (See Exercise 5.8.)

f) Show that $\widehat{(f * g)}(\alpha) = \hat{f}(\alpha)\hat{g}(\alpha)$. (See Exercise 5.47.)

6.8   Let $g \in L^1(\mathbb{R}^n, \mathcal{L}, l)$, and define the mapping

$$T : L^1(\mathbb{R}^n, \mathcal{L}, l) \to L^1(\mathbb{R}^n, \mathcal{L}, l)$$

by $T(f) = f * g$. Prove that if $f_n \to f$ in the L^1-norm, then $T(f_n) \to T(f)$ in the L^1-norm. That is, prove that T is a continuous mapping.

6.9   Let f and g be in $L^1(\mathbb{R})$, and suppose also that g is *essentially bounded*: for some $M \in \mathbb{R}$ we have $|g(x)| \leq M$ for *almost* all $x \in \mathbb{R}$.

a) Prove that $f * g(x)$ is a continuous real-valued function defined for all $x \in \mathbb{R}$. That is, show that

$$|f * g(x) - f * g(x_0)| \to 0$$

as $x \to x_0$.

---

[59]The continuous image, or even the homeomorphic image, of a measurable set need not be measurable. See Exercise 7.13.c.

b) Let

$$f(x) = \frac{1}{\sqrt{|x|}} 1_{[-1,0)\cup(0,1]}(x)$$

for all x. Show that $f \in L^1(\mathbb{R})$ but $f * f$ is not continuous at $x = 0$.

6.10 Suppose $A \in \mathfrak{L}$, the family of Lebesgue measurable subsets of the real line, and $B = -A$. Suppose $0 < l(A) < \infty$. Let $f = 1_A$ and let $g = 1_B$.

a) Prove that $g * f(0) > 0$. (Hint: See Exercise 5.43.)

b) Use part (a) to prove Steinhaus's theorem: There exists an interval

$$(-\delta, \delta) \subseteq A - A = \{x - z \mid x \in A, \ z \in A\}.$$

(Hint: Compare with Exercise 3.21, which called for a different proof.)

6.11 Suppose A and B are measurable subsets of the real line, and suppose that $l(A)l(B) > 0$.

a) Use the convolution $1_{-A} * 1_B$, and Fubini's theorem, to prove that there is a measurable set C of positive measure such that $l\left((A + x) \cap B\right) > 0$ for all $x \in C$.

b) Prove that $(B - A) \cap \mathbb{Q} \neq \emptyset$. That is, prove that the set of differences between elements of A and elements of B must include a rational number.

6.12 Let $f : X \to \mathbb{R}$ be a measurable function on the complete finite measure space $(X, \mathfrak{A}, \mu)$. Suppose $g(x, y) = f(x) - f(y)$ is integrable on $X \times X$. Show that f is integrable on X and calculate the numerical value of $\int_{X \times X} g \, d(\mu \times \mu)$.

6.13 Suppose $(X, \mathfrak{A}, \mu)$ and $(Y, \mathfrak{B}, \nu)$ are both σ-finite complete measure spaces. Suppose $f \in L^1(X)$ and $g \in L^1(Y)$. Define

$$h(x, y) = f(x)g(y),$$

and prove that $h \in L^1(X \times Y, \mathfrak{A} \otimes \mathfrak{B}, \mu \times \nu)$.

6.14 We investigate what is called the *essential uniqueness* of translation-invariant measures.

a) Let $(\mathbb{R}^n, \mathcal{L}, l)$ be the standard Euclidean measure space with Lebesgue measure l defined on the σ-field of Lebesgue measurable sets. By Exercise 3.24 we know that l is *translation-invariant*. Suppose that μ is any *other* σ-finite measure defined and translation-invariant on $\mathcal{L}$. Use Fubini's theorem to prove that $\mu = cl$ for some constant c. This is called the *essential uniqueness* of translation-invariant measure.

(Hint: To prove this with Fubini's theorem, let $E \in \mathcal{L}$ be any set of finite measure, let Q be the *unit cube*, and write

$$\mu(E) = \int_{\mathbb{R}^n} 1_Q(y) \, dl(y) \int_{\mathbb{R}^n} 1_E(x) \, d\mu(x).$$

Then write this as a double integral over the product space $\mathbb{R}^{2n}$, and play with the translation invariance of both measures. This proof is modeled on the proof of a more general case published by Shizuo Kakutani [15].)

b) Let $\nu(E)$ be the number of elements in of E—meaning $\nu(E)$ is either finite or simply ∞—for each $E \in \mathcal{L}$. That is, ν is $[0, \infty]$-valued. Is ν translation-invariant? Is ν a constant multiple of l? Do we have a counterexample to the essential uniqueness of translation-invariant measure on $\mathbb{R}^n$?

6.15   Let f be a real-valued function on $\mathbb{R} \times \mathbb{R}$. Suppose that $f(\cdot, y)$ is continuous in the first variable for each fixed y and that $f(x, \cdot)$ is Lebesgue measurable in the second variable for each fixed x. Prove that f is measurable as a function on the plane. (Hint: Express f as a pointwise limit of measurable functions on the plane.)

6.16   Suppose that $(X, \mathfrak{A}, \mu)$ is a complete σ-finite measure space, and let f be a real-valued integrable function in $L^1(X, \mathfrak{A}, \mu)$. Let l denote Lebesgue measure on the real line. Apply Fubini's theorem to the space $X \times \mathbb{R}$ to prove that

$$\int_X f \, d\mu(x) = \int_{\mathbb{R}} \mu\left(f^{-1}(\alpha, \infty)\right) - \mu\left(f^{-1}(-\infty, -\alpha)\right) \, dl(\alpha).$$

The use of a powerful tool such as Fubini's theorem can produce serious errors if the tool is applied in cases that do not satisfy the hypotheses of the theorem. Here are some examples.

6.17   Let $X = Y = \mathbb{N}$ the set of all natural numbers, and let $\mathfrak{A} = \mathfrak{P}(X)$, the power set of the set of natural numbers. Let $\mu = \nu$ be the ordinary *counting measure* on $\mathfrak{A}$, as defined in Exercises 2.15 and 5.9.

a) Show that $\mu \times \nu$ is counting measure on the power set of $\mathbb{N} \times \mathbb{N}$ and that $\mu \times \nu$ is σ-finite.

b) Define

$$f(x, y) = \begin{cases} 2 - 2^{-x} & \text{if } x = y, \\ -2 + 2^{-x} & \text{if } x = y + 1, \\ 0 & \text{if } x \notin \{y, y+1\}. \end{cases}$$

Show that

$$\int_X \int_Y f(x, y) \, d\nu(y) \, d\mu(x) \neq \int_Y \int_X f(x, y) \, d\mu(x) \, d\nu(y),$$

and explain why this does not violate Fubini's theorem.

6.18   Give an alternative solution for Exercise 5.29 by interpreting that exercise in terms of Fubini's theorem applied to a product in which one of the factors is the counting measure.

6.19   For $x \in \mathbb{R}^1$ and $t > 0$, let

$$f(x, t) = \frac{1}{\sqrt{2\pi t}} e^{-\left(\frac{x^2}{2t}\right)}.$$

It is well known that for each $t > 0$, $\int_{-\infty}^{\infty} f(x, t) \, dx = 1$. It is also known that

$$2\frac{\partial f}{\partial t} = \frac{\partial^2 f}{\partial x^2}.$$

If $g(x, t) = \dfrac{\partial f}{\partial t}$, prove or disprove:

$$\int_{-\infty}^{x} \int_{s}^{\infty} g(x, t)\, dt\, dx \neq \int_{s}^{x} \int_{-\infty}^{\infty} g(x, t)\, dx\, dt.$$

What is the relevance of this example to Fubini theorem?

6.20 Let $X = Y = [0, 1]$. Let μ be Lebesgue measure on X and let λ be counting measure on Y. Let

$$f(x, y) = \begin{cases} 1 & \text{if } x = y \\ 0 & \text{if } x \neq y. \end{cases}$$

Show that

$$\int_{X} \int_{Y} f(x, y)\, d\mu(x)\, d\lambda(y) \neq \int_{Y} \int_{X} f(x, y)\, d\lambda(y)\, d\mu(x)$$

and explain why this does not violate Fubini's theorem.

6.3 COMPARISON OF LEBESGUE AND RIEMANN INTEGRALS

Riemann integration corresponds to the concept of Jordan measure in a manner that is similar (but not identical) to the correspondence between the Lebesgue integral and Lebesgue measure. Although it is possible for an unbounded function to be Lebesgue integrable, this cannot occur with proper Riemann integration. Moreover, proper Riemann integrals are defined only for functions with a bounded domain D. Since a bounded domain D can always be contained in a rectangular block with edges parallel to the axes, and since we can let f be identically zero on the part of the block that is outside D, we will assume that f is defined on such a block. We will denote such a block by the suggestive notation $[\mathbf{a}, \mathbf{b}]$, with the understanding that $\mathbf{a} = (a_1, \ldots, a_n) \in \mathbb{R}^n$ and $\mathbf{b} = (b_1, \ldots, b_n) \in \mathbb{R}^n$. Then the symbol we have chosen for a block has the form

$$[\mathbf{a}, \mathbf{b}] = \prod_{i=1}^{n} [a_i, b_i],$$

a Cartesian product of closed, finite intervals.

Let f be any bounded real-valued function on $[\mathbf{a}, \mathbf{b}]$. Since $f = f^+ - f^-$, a difference between two nonnegative functions, it will suffice to deal with the Riemann integration of nonnegative bounded functions f.[60]

Let Δ denote a partition of $[\mathbf{a}, \mathbf{b}]$ into the union of N rectangular blocks, the *interiors* of which are mutually disjoint:

$$\Delta = \{[\mathbf{x}_i, \mathbf{y}_i] \mid i = 1, \ldots, N\}.$$

---

[60]We assume the reader knows from advanced calculus that the positive and negative parts of a Riemann integrable function must be Riemann integrable. See [20].

Let Δx_i denote the *volume* of the box $[\mathbf{x}_i, \mathbf{y}_i]$. On each of the N blocks $[\mathbf{x}_i, \mathbf{y}_i]$ we let $m_i = \inf\{f(x) \mid x \in [\mathbf{x}_i, \mathbf{y}_i]\}$ and $M_i = \sup\{f(x) \mid x \in [\mathbf{x}_i, \mathbf{y}_i]\}$. We form the *lower* and *upper sums*

$$s(\Delta) = \sum_1^N m_i \Delta x_i$$

and

$$S(\Delta) = \sum_1^N M_i \Delta x_i.$$

Then we define the *lower* and *upper Riemann integrals* by

$$\underline{\int_a^b} f(x)\, dx = \sup_\Delta s(\Delta),$$

which is a supremum over all possible finite partitions Δ of $[\mathbf{a}, \mathbf{b}]$, and

$$\overline{\int_a^b} f(x)\, dx = \inf_\Delta S(\Delta).$$

Definition 6.3.1 A bounded real-valued function f defined on $[\mathbf{a}, \mathbf{b}]$ is called *Riemann integrable* if and only if

$$\overline{\int_a^b} f(x)\, dx = \underline{\int_a^b} f(x)\, dx.$$

In the case of equality, this value is called $\int_a^b f(x)\, dx$.[61]

Theorem 6.3.1 *(Lebesgue's theorem) A bounded real-valued function f on $[\mathbf{a}, \mathbf{b}]$ is Riemann integrable if and only if the set of points x at which f is not continuous is a Lebesgue null set.*

Proof: Without loss of generality, we can suppose that f is a nonnegative bounded function on $[\mathbf{a}, \mathbf{b}]$, since $f = f^+ - f^-$. Note that f is Riemann integrable if and only if the same is true of both f^+ and f^-, and a similar statement applies for continuity at a point x. Let

$$C(f) = \{(x, y) \mid 0 \leqslant y \leqslant f(x)\},$$

---

[61] The reader who wishes to learn more about the Riemann integral in $\mathbb{R}^n$ can consult an advanced calculus text, such as [20].

the region between the graph of f and the block $[\mathbf{a}, \mathbf{b}]$ in the $\mathbb{R}^n$. Observe that the Jordan inner and outer measure of $C(f)$ correspond as follows to the lower and upper Riemann integrals of f:

$$\underline{v}(C(f)) \;=\; \underline{\int_{\mathbf{a}}^{\mathbf{b}}} f(x)\,dx \;=\; \sup_{0 \leqslant g \leqslant f} \int_{\mathbf{a}}^{\mathbf{b}} g(x)\,dx, \qquad (6.8)$$

$$\overline{v}(C(f)) \;=\; \overline{\int_{\mathbf{a}}^{\mathbf{b}}} f(x)\,dx \;=\; \inf_{f \leqslant g} \int_{\mathbf{a}}^{\mathbf{b}} g(x)\,dx, \qquad (6.9)$$

where g varies over the *step functions*.[62] It follows that f is Riemann integrable if and only if $C(f)$ is Jordan measurable, and the latter condition is equivalent to the *boundary* $\partial C(f)$ being a Jordan null set, according to Theorem 3.6.3. Since the boundary is a closed set, this is equivalent to $\partial C(f)$ being a Lebesgue null set.

On the other hand, by Fubini's theorem, $\partial C(f)$ is a Lebesgue null set in the plane if and only if the x-section $_x\partial C(f)$ has linear Lebesgue measure equal to zero for almost all $x \in [\mathbf{a}, \mathbf{b}]$.

Suppose that the point $x = p$ is a point of discontinuity of f. That is, suppose it is false that $\lim_{x \to p} f(x) = f(p)$. Since f is bounded, the Bolzano-Weierstrass theorem can be used to establish that there is in $[\mathbf{a}, \mathbf{b}]$ a sequence $x_n \to p$ such that $f(x_n) \to y_1 \neq f(p)$. Denote $f(p) = y_2$. Suppose that $y_1 < y_2$. (The argument would be nearly identical with the opposite inequality.) The reader should show that

$$_p\partial C(f) \supseteq \{(p, y) \mid y_1 \leqslant y \leqslant y_2\},$$

so that $_p\partial C(f)$ has strictly positive linear Lebesgue measure. Moreover, if f is continuous at p, then $_p\partial C(f)$ contains at most two points and is a Lebesgue null set. See Exercise 6.24.   ∎

■ EXAMPLE 6.1

Let $f : [0, 1] \times [0, 1] \to \mathbb{R}$ be defined by

$$f(x_1, x_2) = \begin{cases} \frac{1}{2} \sin \frac{\pi}{x_1 x_2} & \text{if } 0 < x_i \leqslant 1, i = 1, 2, \\ 0 & \text{if } x_1 x_2 = 0. \end{cases}$$

We claim that f is Riemann integrable on the closed rectangular box

$$[(0, 0), (1, 1)] = [0, 1]^2.$$

(See Figure 6.1)[63]. In fact, f is bounded and continuous except at the points on the two axes:

$$S = \{(x_1, x_2) \mid x_1 x_2 = 0\}.$$

---

[62] In this context, by a step function we mean a finite linear combination of indicator functions of rectangular boxes.

[63] This illustration is from [20]

Since S is a Lebesgue null set in the Lebesgue measure on $\mathbb{R}^2$, Lebesgue's theorem implies the integrability of f.

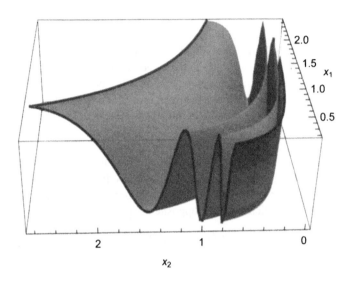

Figure 6.1 $f(\mathbf{x}) = \frac{1}{2} \sin \frac{\pi}{x_1 x_2}$.

We remark that a Fubini theorem for the Riemann integral is much less general and more cumbersome in its statement than is the case for the Lebesgue integral. One reason for this is that one cannot be assured of the existence of the iterated integrals, and far fewer functions are Riemann integrable than Lebesgue integrable.

EXERCISES

6.21   Suppose $f : [0, 1] \to \mathbb{R}$ is given by $f = 1_{\mathbb{Q} \cap [0,1]}$. Prove that f is not Riemann integrable by applying Lebesgue's theorem.

6.22   Let $f : [0, 1] \to \mathbb{R}$ by letting

$$f = 1_{\left\{ \frac{1}{n} \mid n \in \mathbb{N} \right\}}.$$

Prove by using Lebesgue's theorem that f is Riemann integrable.

6.23   Let $f : [0, 1] \to \mathbb{R}$ by letting $f = 1_C$, where C is the Cantor set. (See Exercise 3.11.) Prove by using Lebesgue's theorem that f is Riemann integrable.

6.24   Let $f : [a, b] \to \mathbb{R}$ be a nonnegative function defined on a block in Euclidean space. Define

$$C(f) = \{(\mathbf{x}, y) \mid 0 \leqslant y \leqslant f(\mathbf{x})\},$$

as in the proof of Theorem 6.3.1. Prove by the following steps that f is continuous x if and only if $l\left(\,_\mathbf{x}\partial C(f)\right) = 0$.

a) Show that if f is continuous at $\mathbf{p}$, then $_\mathbf{p}\partial C(f)$ consists of at most two points.

b) Let $p \in [a, b]$. Show that if it is false that $\lim_{\mathbf{x}\to\mathbf{p}} f(\mathbf{x})$ exists and equals $f(\mathbf{p})$, then $_\mathbf{p}\partial C(f)$ contains an interval of positive length.

CHAPTER 7

FUNCTIONS OF A REAL VARIABLE

The purpose of this chapter is to consider the relationship between Lebesgue integration and differentiation for functions of a real variable. Thus this chapter is concerned with the adaptation to the Lebesgue integral of what is called the *fundamental theorem of the calculus* for a single real variable. We will show that for each Lebesgue integrable function f, the indefinite integral $F(x) = \int_a^x f \, dl$ has bounded variation, is differentiable almost everywhere, and $F'(x) = f(x)$ almost everywhere.

7.1 FUNCTIONS OF BOUNDED VARIATION

Definition 7.1.1 A function $f : I \to \mathbb{R}$ is said to be of bounded variation on an interval I, written as $f \in \mathcal{BV}(I)$, provided that there exists a real number $M > 0$ such that the *variation*

$$v(\Delta) = \sum_{i=1}^{n} |f(t_i) - f(t_{i-1})| \leqslant M$$

for every partition

$$\Delta = \{t_0 \leqslant t_1 \leqslant \ldots \leqslant t_n\} \subset I$$

Measure and Integration: A Concise Introduction to Real Analysis. By Leonard F. Richardson
Copyright © 2009 John Wiley & Sons, Inc.

of I by finitely many points $t_0, \ldots, t_n$.[64] If $I = [a, b]$ is a closed finite interval, we require that the partition Δ have $t_0 = a$ and $t_n = b$.

EXERCISE

7.1   Prove that $BV(I)$ is a vector space. That is, prove that if both f and g are in $BV(I)$, then the same is true for $af + g$ for each $a \in \mathbb{R}$.

Define the *positive variation* and the *negative variation* corresponding to a partition Δ by

$$p(\Delta) = \sum_1^n (f(t_i) - f(t_{i-1}))^+, \text{ and} \tag{7.1}$$

$$n(\Delta) = \sum_1^n (f(t_i) - f(t_{i-1}))^-,$$

where, in general, x^+ means the positive part of the real number x, and it serves only to replace x by zero if it is negative. The superscript x^- means the negative part of x, and it serves to replace the value of x by zero if it is positive and by its *absolute value* if it is negative.

Definition 7.1.2 If $f : [a, b] \to \mathbb{R}$ has bounded variation, and if $x \in [a, b]$, then f is still of bounded variation on the subinterval $[a, x]$. Let Δ denote an arbitrary partition of $[a, x]$, and define $p(\Delta)$ and $n(\Delta)$ as in Equations (7.1). Define

$$p(x) = \sup_\Delta p(\Delta),$$

$$n(x) = \sup_\Delta n(\Delta), \text{ and}$$

$$v(x) = \sup_\Delta v(\Delta),$$

each of which is bounded above by M, as defined in Definition 7.1.1.

Theorem 7.1.1 *A function* $f : [a, b] \to \mathbb{R}$ *lies in* $BV[a, b]$ *if and only if* f *can be represented as*

$$f = f_1 - f_2,$$

where f_1 *and* f_2 *are both monotonically increasing functions. Moreover, if we have* $f \in BV[a, b]$, *and if* v, n, *and* p *are as in Definition 7.1.2, then*

$$f(x) - f(a) = p(x) - n(x), \text{ and} \tag{7.2}$$

$$v(x) = p(x) + n(x) \tag{7.3}$$

---

[64]Note that M is required to be independent of both the choice of $n \in \mathbb{N}$ and the choice of partition Δ. There are uncountably many choices of Δ for any one choice of $n \geq 2$ and a fixed interval $[a, b]$.

for all $x \in [a, b]$.

Proof: Sufficiency is easy to establish. If f is any *monotone* function on $[a, b]$, whether it is increasing or decreasing, then

$$v(\Delta) = |f(b) - f(a)|$$

for all Δ. Thus f_1 and f_2 have bounded variation, and we apply Exercise 7.1.

We turn next to the proof of necessity. So we suppose that $f \in \mathcal{BV}[a, b]$. Let $x \in [a, b]$, and let Δ be any partition of $[a, x]$.

It is easy to see that

$$v(x) = \sup_{\Delta} (p(\Delta) + n(\Delta))$$

$$\overset{(i)}{\leqslant} \sup_{\Delta} p(\Delta) + \sup_{\Delta} n(\Delta) \tag{7.4}$$

$$= p(x) + n(x)$$

from the definitions. We need to prove equality in Inequality (i) of Equation (7.4). For $\epsilon > 0$ and some suitable partitions Δ_1 and Δ_2 of $[a, x]$, we have

$$p(x) - p(\Delta_1) < \frac{\epsilon}{2}, \text{ and}$$

$$n(x) - n(\Delta_2) < \frac{\epsilon}{2}.$$

Let $\Delta = \Delta_1 \cup \Delta_2$. We claim that

$$p(\Delta) \geqslant p(\Delta_1), \text{ and} \tag{7.5}$$

$$n(\Delta) \geqslant n(\Delta_2).$$

The reader should check that this is so because partitioning an interval $[x_{i-1}, x_i]$ by means of a point x of that interval forces

$$\left(f(x_i) - f(x)\right)^+ + \left(f(x) - f(x_{i-1})\right)^+ \geqslant \left(f(x_i) - f(x_{i-1})\right)^+$$

and

$$\left(f(x_i) - f(x)\right)^- + \left(f(x) - f(x_{i-1})\right)^- \geqslant \left(f(x_i) - f(x_{i-1})\right)^-.$$

It is simply a matter of checking two cases: Either $f(x)$ lies between $f(x_i)$ and $f(x_{i-1})$, or it does not.

It follows from Inequalities (7.5) that

$$v(x) \geqslant v(\Delta) = p(\Delta) + n(\Delta)$$

$$> p(x) - \frac{\epsilon}{2} + n(x) - \frac{\epsilon}{2}$$

$$= p(x) + n(x) - \epsilon$$

for each $\epsilon > 0$. This establishes that

$$v(x) = p(x) + n(x). \tag{7.6}$$

It is clear that the three functions $p, n,$ and v are all monotone increasing, since any partition Δ of $[a, x]$ can be extended to a partition $\Delta \cup \{x'\}$ of $[a, x']$ for $x' > x$.

Note that

$$p(\Delta) - n(\Delta) = f(x) - f(a)$$

for each partition Δ of $[a, x]$. Since this implies that

$$p(\Delta) + f(a) = n(\Delta) + f(x),$$

it follows that

$$\sup_{\Delta} p(\Delta) + f(a) = \sup_{\Delta'} n(\Delta') + f(x).$$

Thus

$$p(x) - n(x) = f(x) - f(a).$$

We have shown that f can be expressed as the difference of two monotone increasing functions having the additional special property of satisfying Equation (7.6). ∎

Remark 7.1.1 We note that the representation $f(x) - f(a) = p(x) - n(x)$ in Theorem 7.1.1 is not unique. In fact, we could take any monotone increasing function $h(x)$ and write

$$f(x) - f(a) = [p(x) + h(x)] - [n(x) + h(x)].$$

However, the reader will prove in Exercise 7.2 that only one such decomposition satisfies the requirement that

$$v(x) = p(x) + n(x).$$

EXERCISES

7.2 Suppose for f, as in Theorem 7.1.1, we had monotone increasing functions f_1 and f_2 such that

$$f(x) - f(a) = p(x) - n(x) = f_1(x) - f_2(x).$$

Observe that $f_1(x) - p(x) = f_2(x) - n(x)$ and prove that

$$p(x) + n(x) \leqslant \big(f_1(x) - f_1(a)\big) + \big(f_2(x) - f_2(a)\big).$$

This establishes that the decomposition in terms of p and n in Theorem 7.1.1 is minimal in the sense that f_1 and f_2 must increase faster than p and n do, and that the difference cancels out, as in Remark 7.1.1.

7.3 If f is continuous on an interval $[a, b]$ and has a bounded derivative in (a, b), show that f is of bounded variation on $[a, b]$. Is the boundedness of f' necessary for f to be of bounded variation? Justify your answer.

7.4 Let

$$f_n(x) = \begin{cases} x^n \sin\left(\frac{\pi}{x}\right) & \text{if } 0 < x \leqslant 1, \\ 0 & \text{if } x = 0. \end{cases}$$

We claim that $f_n \in \mathcal{BV}[0, 1]$ if $n \geqslant 2$ but it is not in this space if $n = 1$. See Figures 7.1 and 7.2.[65]

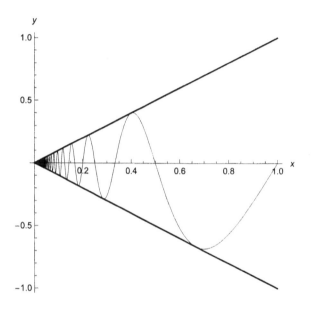

Figure 7.1 $f(x) = x \sin\left(\frac{\pi}{x}\right)$, with envelope $u(x) = x$, $l(x) = -x$.

7.5 If both f and g are in $\mathcal{BV}[a, b]$, prove that $fg \in \mathcal{BV}[a, b]$. (Caution: The product of two monotone functions need not be monotone.)

Remark 7.1.2 Let $f : [a, b] \to \mathbb{R}$ be Lebesgue integrable, which is equivalent to f^+ and f^- being Lebesgue integrable. The *indefinite integral* of f is given by

$$F(x) = \int_a^x f(t)\, dl(t)$$
$$= \int_a^x f^+(t)\, dl(t) - \int_a^x f^-(t)\, dl(t),$$

which is a difference of two monotone increasing functions, so that F lies in $\mathcal{BV}[a, b]$.

---

[65]These two figures are from [20].

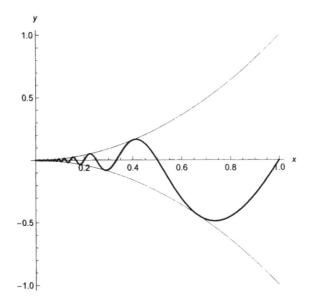

Figure 7.2 $f(x) = x^2 \sin\left(\frac{\pi}{x}\right)$, with envelope $u(x) = x^2$, $l(x) = -x^2$.

Thus, although $f \in L^1(\mathbb{R}, \mathfrak{L}, l)$ need not be Riemann integrable, the *indefinite integral* of f has bounded variation and is therefore Riemann integrable, being a difference of two monotone functions. These observations are significant for the theory of functions of a real variable, and they lead us to the Fundamental Theorem of Calculus for the Lebesgue integral in the next section. First, in preparation, the reader should solve the following easy exercise.

EXERCISE

7.6   Give an example of a Lebesgue integrable function f on $[0, 1]$ for which

$$\frac{d}{dx} \int_0^x f(t)\, dl(t)$$

exists for all $x \in [0, 1]$ but fails to be equal to $f(x)$ for $x \in [0, 1] \cap \mathbb{Q}$.

7.2 A FUNDAMENTAL THEOREM FOR THE LEBESGUE INTEGRAL

Theorem 7.2.1 *Let f be any Lebesgue integrable function on $[a, b]$, and define the* indefinite integral $F(x)$ *by*

$$F(x) = \int_a^x f(t)\, dl(t)$$

for all $x \in [a, b]$. Then the derivative $F'(x)$ exists for almost all x, and $F'(x) = f(x)$ almost everywhere.

Proof: We know that $F \in \mathcal{BV}[a, b]$ by Remark 7.1.2. The differentiability of $F(x)$ for almost all x will follow therefore from Lebesgue's theorem (7.3.1), which we will prove in the next section.[66] For now we will assume Lebesgue's theorem and prove the remaining conclusions of Theorem 7.2.1. It will suffice to give a proof for $f(x) \geqslant 0$ for all $x \in [a, b]$, since in general

$$ f = f^{+} - f^{-}, $$

a difference of two positive integrable functions. In summary, we are assuming that $F'(x)$ exists for almost all x, and we must prove that $F'(x) = f(x)$ almost everywhere.

We consider first the case in which the function f is bounded:

$$ 0 \leqslant f(t) \leqslant M \in \mathbb{R} $$

for all $t \in [a, b]$. We observe that the indefinite integral, $F(x)$, is a monotone increasing *continuous* function of x since f is both nonnegative and integrable. Moreover, the difference quotients

$$ \frac{F(x + h_n) - F(x)}{h_n} \to F'(x) $$

for almost all x, independent of the choice of sequence $h_n \to 0$. Thus F' is a measurable nonnegative function, defined almost everywhere. And it is easy to calculate that

$$ \frac{F(x + h_n) - F(x)}{h_n} \leqslant M, $$

which is an integrable constant function on $[a, b]$. This is the place where the boundedness of f is helpful—because M is an integrable constant on each finite interval. By Lebesgue Dominated Convergence (Theorem 5.3.1), we know for each interval $[c, d] \subset (a, b)$ that

$$ \int_{c}^{d} \frac{F(x + h_n) - F(x)}{h_n} \, dl(x) \to \int_{c}^{d} F'(x) \, dl(x) $$

as $n \to \infty$. However, the left-hand side can be written as

$$ \frac{1}{h_n} \left(\int_{d}^{d+h_n} F(x) \, dl(x) - \int_{c}^{c+h_n} F(x) \, dl(x) \right) \to F(d) - F(c) $$

---

[66]This is not the same Lebesgue theorem that we saw in the preceding chapter, which classified the Riemann integrable functions. Both theorems are commonly called *Lebesgue's theorem*.

by the Mean Value Theorem for integrals applied to the *continuous* integrand F. It follows from the uniqueness of limits that

$$\int_c^d f(x)\, dl(x) = F(d) - F(c) = \int_c^d F'(x)\, dl(x).$$

The latter conclusion can be rewritten as

$$\int_c^d F'(x) - f(x)\, dl(x) = 0$$

for all $[c, d] \subset (a, b)$. It follows that $F'(x) = f(x)$ almost everywhere.[67]

The next case permits f to be an *unbounded* nonnegative integrable function. Define the *truncation*

$$f_n(t) = (f \wedge n)(t) = \min\big(f(t), n\big)$$

for each $n \in \mathbb{N}$ and for each $t \in \mathbb{R}$. Let

$$F_n(x) = \int_a^x f_n\, dl.$$

By the first case, $F_n'(x) = f_n(x)$ for almost all x. Since the monotone increasing sequence $f_n \to f$, we have the increasing sequence $F_n \to F$ by Monotone Convergence (Theorem 5.4.1). And F, like F_n, is a monotone increasing function of x. In fact, for each fixed n, consider the difference

$$F(x) - F_n(x) = \int_a^x f - f_n\, dl,$$

which is an increasing function of x. By Lebesgue's theorem (7.3.1), we have $F - F_n$ differentiable almost everywhere. And

$$\begin{aligned}
F'(x) &= (F - F_n)'(x) + F_n'(x) \\
&= (F - F_n)'(x) + f_n(x) \\
&\geqslant f_n(x)
\end{aligned}$$

for almost all x, since the derivative of an increasing function must be nonnegative wherever the derivative exists.

Thus

$$F'(x) - f(x) \geqslant 0 \tag{7.7}$$

---

[67]The reader may find it interesting to compare this method of proving that a function is zero almost everywhere with Exercise 3.8. That exercise implies that no set could have the property of comprising exactly half of each interval in measure. Thus, for example, it is not possible for a measurable function to be alternately 1 and -1 on half of each interval, making the integral zero without f being zero almost everywhere. These observations are not needed, however, to justify the method we have just used to prove that $F' = f$ almost everywhere.

almost everywhere. Also,

$$
\begin{aligned}
\int_c^d F'(x) - f(x)\, dl(x) &= \int_c^d \lim_{h_n \to 0} \frac{F(x+h_n) - F(x)}{h_n}\, dl(x) - \int_c^d f(x)\, dl(x) \\
&\leqslant \lim_{h_n \to 0} \int_c^d \frac{F(x+h_n) - F(x)}{h_n}\, dl(x) - \int_c^d f(x)\, dl(x) \\
&= F(d) - F(c) - \int_c^d f(x)\, dl(x) \\
&= 0,
\end{aligned}
$$

where the *inequality* above comes from Fatou's lemma $\big($Theorem (5.4.3)$\big)$. We have used also the fact that F is continuous because f is integrable. It follows that $F'(x) \leqslant f(x)$ for almost all x. Because of Equation (7.7), $F'(x) = f(x)$ almost everywhere. ∎

EXERCISE

7.7 Let $\phi : \mathbb{R} \to \mathbb{R}$ be a *measurable homomorphism* of the additive group of real numbers. That is, ϕ is measurable, and

$$\phi(x+y) \equiv \phi(x) + \phi(y). \tag{7.8}$$

Suppose also that ϕ is *locally integrable*,[68] meaning that for each $p \in \mathbb{R}$ there exists $r > 0$ such that

$$\int_{p-r}^{p+r} \phi \, dl$$

exists. Use Theorem 7.2.1 to prove that ϕ is continuous and also that ϕ is a *linear* mapping of $\mathbb{R}$ to itself.[69]

7.3 LEBESGUE'S THEOREM AND VITALI'S COVERING THEOREM

In this section we will prove the theorem of Lebesgue that we have used already.

Theorem 7.3.1 (Lebesgue) *Let $f \in \mathcal{BV}[a, b]$. Then $f'(x)$ exists and is finite for almost all $x \in (a, b)$.*

---

[68]Local integrability is not necessary for the stated conclusion to be true. Local integrability does permit, however, an easy proof using Theorem 7.2.1. See Exercise 4.14 for a hint for a fairly simple proof based on Lusin's theorem, without assuming local integrability. Thus all measurable homomorphisms of the additive group of real numbers to itself must be continuous. Exercise 7.13.c shows that measurability is *not* a topological property. It is interesting that measurability combined with the homomorphism property, which is also not topological, implies continuity.

[69]Cauchy proved that every continuous function satisfying Equation (7.8) must be linear. Georg Hamel showed in [11] that there exist solutions of that functional equation that are not continuous and hence not linear. It was in [11] that Hamel introduced the *Hamel basis* for the set of real numbers.

One could extend the theorem slightly by considering the right-hand derivative at a and the left-hand derivative at b, but this has no effect upon the measure-theoretic claim. By Theorem 7.1.1 it suffices to prove the theorem for f monotonically increasing. In order to prove the theorem of Lebesgue, we must prove first another famous theorem.

Theorem 7.3.2 (Vitali's Covering Theorem) *Let A be a subset of a finite open interval (a, b). Let $\mathcal{I} = \{I\}$ be a family of closed subintervals I of strictly positive length in (a, b) with the following Vitali Property:*

- *For each $x \in A$, and for each $\delta > 0$, there exists $I \in \mathcal{I}$ such that $x \in I$[70] and $|I| < \delta$.*

Then there exists a countable set of mutually disjoint intervals $I_n \in \mathcal{I}$ that cover A up to a null set, meaning that

$$l\left(A \setminus \bigcup_{n \in \mathbb{N}} \dot{I_n}\right) = 0.$$

We will denote the property of a set A being covered *up to a null set* as

$$A \overset{\circ}{\subseteq} \bigcup_{n \in \mathbb{N}} \dot{I_n}.$$

We remark that in Vitali's theorem the set $A \subseteq (a, b)$ does not have to be measurable, although the conclusion states that the coverage of A is up to a measurable set of measure zero.

Proof: The following notation will be convenient: If $I \subset (a, b)$, denote

$$I^c = (a, b) \setminus I.$$

We proceed to the construction of the sequence of intervals $I_n \in \mathcal{I}$ that cover A up to a null set as follows. The process described in the display below may proceed without end or it may be forced to terminate.[71] We let

---

[70]It is acceptable for x to be an endpoint of the closed interval I.
[71]It is interesting to note that other than requiring $\mathcal{I}$ to be a Vitali covering of A, the recipe for the construction of the sequence I_n proceeds without reference to A.

$$\alpha_1 = \sup\left\{|I| \,\big|\, I \in \mathcal{I}\right\}, \text{ and}$$

$$\alpha_1 > 0 \implies \exists I_1, \; |I_1| > \frac{\alpha_1}{2};$$

$$\alpha_2 = \sup\left\{|I| \,\big|\, I \in \mathcal{I}, \, I \subset I_1^c\right\}, \text{ and}$$

$$\alpha_2 > 0 \implies \exists I_2 \subset I_1^c, \; |I_2| > \frac{\alpha_2}{2};$$

$$\vdots = \vdots$$

$$\alpha_n = \sup\left\{|I| \,\big|\, I \in \mathcal{I}, \, I \subset \bigcap_{j<n} I_j^c\right\}, \text{ and}$$

$$\alpha_n > 0 \implies \exists I_n \subset \bigcap_{j<n} I_j^c, \; |I_n| > \frac{\alpha_n}{2}.$$

Consider first the possibility that this process terminates in $n-1$ steps, meaning that no I_n of positive length is disjoint from all those already selected. We claim in this case that

$$A \subseteq \bigcup_{j<n} I_j.$$

If it were the case that A is not covered by $\bigcup_{j<n} I_j$, then there would exist $x \in A$ such that x lies in the *complement* of $\bigcup_{j<n} I_j$, which is *open*. Then there must exist an I from the Vitali covering that contains x but has length too short to intersect the union of the selected intervals. This is a contradiction. Thus termination of the process would imply that A is covered entirely by a finite sequence of intervals from the covering.

So consider the remaining and main case, that the process does not terminate. Since the length of (a, b) is finite and the intervals I_n selected as above must be disjoint, it follows that $|I_n| \to 0$ and $\alpha_n \to 0$ also. We claim that

$$A \overset{\circ}{\subseteq} \dot{\bigcup_{n \in \mathbb{N}}} I_n.$$

If this were false, then we would have the strictly positive number

$$\eta = l^*\left(A \setminus \dot{\bigcup_{n \in \mathbb{N}}} I_n\right) > 0.$$

Let J_n be defined as the closed interval with the same midpoint as I_n but with[72]

$$|J_n| = 5|I_n|.$$

---

[72]The use of the number 5 in this theorem is so distinctive that some authors refer to this theorem as the *Vitali Five Theorem*.

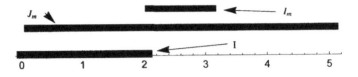

Figure 7.3   Top to bottom: $I_m, J_m, I; |I| > 2|I_m|$.

Then $\sum_{1}^{\infty} |J_i| < \infty$, although it is *not necessary* to have $J_n \subset (a, b)$.

There exists $N \in \mathbb{N}$ such that $\sum_{N+1}^{\infty} |J_n| < \eta$. Thus

$$A \overset{\circ}{\nsubseteq} \left(\bigcup_{n \leqslant N} I_n \right) \cup \left(\bigcup_{N+1}^{\infty} J_n \right).$$

Thus there exists $x_0 \in A$ such that $x_0 \notin \bigcup_{n \leqslant N} I_n$ and $x_0 \notin J_n$ for any $n \geqslant N + 1$.
We will explain why this yields a contradiction.

By the hypotheses of Vitali's theorem, there exists a sufficiently short interval $I \in \mathcal{I}$ such that $x_0 \in I$ and

$$I \subset \bigcap_{n \leqslant N} I_n^c.$$

Since $\alpha_n \to 0$, there exists n_0 such that $\alpha_{n_0} < |I|$. Hence I is too long to be disjoint from all the intervals I_n with $n < n_0$. Thus

$$I \cap \left(\bigcup_{n < n_0} I_n \right) \neq \varnothing.$$

Let m be the *least* value of n such that $I \cap I_n \neq \varnothing$. It follows that $m \geqslant N + 1$. And since $x_0 \notin J_n$ if $n \geqslant N + 1$, we see that $x_0 \notin J_m$. But $x_0 \in I \backslash J_m$, and yet $I \cap I_m \neq \varnothing$. Since $|J_m| = 5|I_m|$, it follows that

$$|I| > 2|I_m| > 2\frac{\alpha_m}{2} = \alpha_m$$

so that $|I| > \alpha_m$. (See Figure 7.3.) Thus $I \cap I_n \neq \varnothing$ for some $n < m$, contradicting the minimality of the choice of m. This is the contradiction that proves Vitali's theorem. ∎

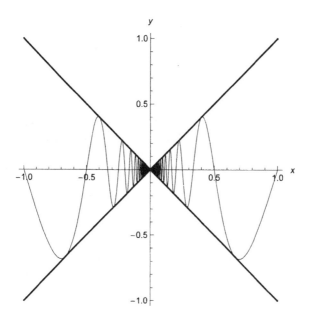

Figure 7.4   Unequal upper and lower derivatives for $f(x) = x \sin\left(\frac{\pi}{x}\right)$.

Definition 7.3.1 We define the upper and lower left- and right-hand derivatives of a function defined in a neighborhood of $x \in \mathbb{R}$ as follows:

$$D^+ f(x) = \limsup_{k \to 0+} \frac{f(x+k) - f(x)}{k} = \lim_{h \to 0+} \sup_{0 < k < h} \left\{ \frac{f(x+k) - f(x)}{k} \right\},$$

$$D_+ f(x) = \liminf_{k \to 0+} \frac{f(x+k) - f(x)}{k} = \lim_{h \to 0+} \inf_{0 < k < h} \left\{ \frac{f(x+k) - f(x)}{k} \right\},$$

$$D^- f(x) = \limsup_{k \to 0-} \frac{f(x+h) - f(x)}{h} = \lim_{h \to 0-} \sup_{h < k < 0} \left\{ \frac{f(x+k) - f(x)}{k} \right\},$$

$$D_- f(x) = \liminf_{k \to 0-} \frac{f(x+k) - f(x)}{k} = \lim_{h \to 0-} \inf_{h < k < 0} \left\{ \frac{f(x+k) - f(x)}{k} \right\}.$$

Recall that the lim sup and lim inf always exist within the extended real number system $\mathbb{R}^* = \mathbb{R} \cup \{\pm\infty\}$. It is not hard to prove that $f'(x)$ exists if and only if the values of all four upper and lower right and left derivatives are equal and real-valued.

EXERCISE

7.8 Let

$$f(x) = \begin{cases} x \sin\left(\frac{\pi}{x}\right) & \text{if } x \neq 0, \\ 0 & \text{if } x = 0. \end{cases}$$

Find all four upper and lower one-sided derivatives of f at $x = 0$. (See Figure 7.4.)

We are ready to prove Theorem 7.3.1.

Proof: Recall that we can assume without loss of generality that f is an increasing function on $[a, b]$. Let

$$\mathcal{D} = \left\{ x \in (a, b) \mid 0 \leqslant D^+ f(x) = D^- f(x) = D_+ f(x) = D_- f(x) < \infty \right\}.$$

We will prove that $l\big((a, b)\backslash\mathcal{D}\big) = 0$. Note that there is no significant loss in omitting the endpoints of the interval $[a, b]$, because a two-point set is a null set. Since the monotone function f lies in $\mathcal{BV}[a, b]$, f is bounded and both $f(a)$ and $f(b)$ are real-valued. We will present most of the work of the proof in the form of two lemmas.

Lemma 7.3.1 *If* $A = \{x \mid a < x < b, D^+ f(x) = \infty\}$, *then A is measurable, and* $l(A) = 0$.

Proof: We are not assuming that A is measurable. However, we will prove that $l^*(A) = 0$, and this will imply that A is a null set. Fix $\beta > 0$, arbitrarily large. If $x \in A$, then there exist values of $h > 0$ as small as we like such that

$$\frac{f(x + h) - f(x)}{h} > \beta.$$

Let

$$\mathcal{I} = \left\{ [c, d] \subset (a, b) \;\middle|\; \left| \frac{f(d) - f(c)}{d - c} \right| > \beta \right\}.$$

Then $\mathcal{I}$ covers A in the sense of Vitali. By Vitali's theorem, there exists a sequence $I_n \in \mathcal{I}$ of *mutually disjoint* intervals such that $A \overset{\circ}{\subseteq} \bigcup_{n \in \mathbb{N}} I_n$. Write

$$I_n = [c_n, d_n].$$

Then

$$f(d_n) - f(c_n) > \beta(d_n - c_n), \quad \text{and}$$

$$f(b) - f(a) \geqslant \sum_1^\infty \big(f(d_n) - f(c_n)\big)$$

$$> \beta \sum_1^\infty |I_n|.$$

Thus

$$\sum_1^\infty |I_n| < \frac{f(b) - f(a)}{\beta},$$

which can be made as small as we like by increasing β. Thus $l^*(A) = 0$.   ∎

Note that because f is monotone increasing, the lower derivative in either direction must be nonnegative. This implies that neither $D_+ f(x)$ nor $D_- f(x)$ can be $-\infty$.

On the other hand, if $D_+ f(x) = \infty$, then the same is true for $D^+ f(x)$, so x belongs to the null set identified in Lemma 7.3.1. For the cases of $D^- f(x)$ and $D_- f(x)$ we let $g(x) = -f(-x)$ on $[-b, -a]$, which interchanges the lower right derivative with the upper left derivative and the lower left with the upper right.

Thus we can restrict our attention without loss of generality to the case in which both one-sided upper and lower derivatives are finite, which we assume henceforth.

Lemma 7.3.2 *Let* $A = \{x \in (a, b) \mid D^+ f(x) > D_- f(x)\}$. *Then* A *is measurable, and* $l(A) = 0$.

Proof: Let $r > s$, with both numbers rational, and let

$$A_{r,s} = \left\{x \in (a, b) \mid D^+ f(x) > r > s > D_- f(x)\right\}.$$

We see that

$$A = \bigcup_{(r,s) \in \mathbb{Q}^2} A_{r,s}$$

is a union of countably many sets. (Here (r, s) denotes an ordered pair of rational numbers, *not* an interval.) Thus it suffices to show that each $A_{r,s}$ is a null set.

For each $I = [c, d] \subset (a, b)$ we define $f(I) = f(d) - f(c)$. Suppose that

$$p = l^*(A_{r,s}) > 0.$$

Note that we *do not assume* that $A_{r,s}$ is measurable. We will deduce a contradiction from the assumption that $p > 0$.

Here is the idea of the proof of this lemma. We will begin by identifying a sequence of mutually disjoint intervals on which the sum of the increments of f from one end to the other is bounded above in terms of s. Then we will find a sequence of other intervals lying within the union of the first sequence on which the sum of the increments of f exceeds the aforementioned bound because r is larger than s. If we do this carefully, it will yield an impossible inequality. Now we proceed with the details of the proof.

If $\epsilon > 0$, then there exists an open set $G \supset A_{r,s}$ such that $l(G) < p + \epsilon$. Let

$$\mathcal{I} = \{I \subset G \mid f(I) < s|I|\}$$

so that $\mathcal{I}$ covers $A_{r,s}$ in the sense of Vitali. By Vitali's theorem there exists a sequence of mutually disjoint intervals $I_k \in \mathcal{I}$ such that

$$A_{r,s} \overset{\circ}{\subseteq} \bigcup_{k \in \mathbb{N}} I_k^\circ,$$

where I° denotes the *interior* of the set I. Here we are using the fact that the set of endpoints of the countably many intervals I_k is a null set, so that the theorem is unaffected if we do not use the set of endpoints of the intervals I_k. Let

$$A'_{r,s} = A_{r,s} \cap \left(\bigcup_{k \in \mathbb{N}} I_k^\circ\right),$$

so that $l\left(A_{r,s}\backslash A'_{r,s}\right) = 0$. The combination of monotonicity and subadditivity of l^* implies that

$$l^*(A'_{r,s}) \leqslant l^*(A_{r,s}) = p \leqslant l^*(A'_{r,s}) + 0,$$

which implies that

$$l^*(A'_{r,s}) = p.$$

Now let $G' = \overset{\cdot}{\underset{k\in\mathbb{N}}{\bigcup}} I_k^{\circ}$, which is an open set. Let

$$\mathcal{J} = \left\{J \subseteq G' \,|\, f(J) > r|J|\right\},$$

where J denotes an interval. Then $\mathcal{J}$ covers $A'_{r,s}$ in the sense of Vitali. Hence there exists a sequence of disjoint intervals $J_k \in \mathcal{J}$ such that

$$A'_{r,s} \overset{\circ}{\subseteq} \bigcup_{k\in\mathbb{N}} J_k.$$

Since f is monotone increasing, $f(I)$ and $f(J)$ must always be nonnegative numbers, and

$$\sum_1^\infty f(I_k) \geqslant \sum_1^\infty f(J_k)$$
$$> rl^*(A'_{r,s})$$
$$= rp.$$

Yet it is true also that

$$\sum_1^\infty f(I_k) < s\sum_1^\infty |I_k|$$
$$< sl(G)$$
$$< s(p+\epsilon).$$

Hence $s(p + \epsilon) > rp$, which implies that

$$\frac{p+\epsilon}{p} > \frac{r}{s} > 1.$$

Since we are assuming that $p > 0$, and because $\epsilon > 0$ can be taken as small as we like, this implies that

$$1 \geqslant \frac{r}{s} > 1,$$

which is a contradiction.   ∎

We are ready to complete the proof of Theorem 7.3.1. If we let

$$g(x) = -f(-x)$$

on $[-b, -a]$, then g is also monotone increasing, and

$$D^+g(-x) = D_-f(x).$$

By applying Lemma 7.3.2 to g, we see that the set of points x at which

$$D_-f(x) > D^+f(x)$$

is also a null set. Thus $D^+f = D_-f$ almost everywhere. Since there are four upper and lower one-sided derivatives at each point, it is necessary to consider the remaining pairs. For the two pairs $(D^+f(x), D_+f(x))$ and $(D^-f(x), D_-f(x))$ we can emulate the proof of Lemma 7.3.2 and then replace f by $g(x) = -f(-x)$, which is still increasing, to reverse the roles of the upper and lower right derivatives. Similar work can be done to cover the pairs $(D^+f(x), D^-f(x))$ and $(D_-f(x), D_+f(x))$. ∎

7.4 ABSOLUTELY CONTINUOUS AND SINGULAR FUNCTIONS

The concept of *absolute continuity* for a real-valued function of a real variable is particularly important when studying the various forms of the Fundamental Theorem of Calculus for the Lebesgue integral. We present the definition of this concept below, following a review of two more elementary concepts of continuity.

Definition 7.4.1 Let $f : [a, b] \to \mathbb{R}$. Then we have the following definitions regarding f.

1. The function f is *continuous* at $x_0 \in [a, b]$ if and only if for each $\epsilon > 0$ there exists $\delta > 0$ such that $x \in [a, b]$, with $|x - x_0| < \delta$, implies that

$$|f(x) - f(x_0)| < \epsilon.$$

2. The function f is *uniformly continuous on* $[a, b]$ if and only if for each $\epsilon > 0$ there exists $\delta > 0$ such that x and y in $[a, b]$, with $|x - y| < \delta$, implies that

$$|f(x) - f(y)| < \epsilon.$$

3. The function f is *absolutely continuous on* $[a, b]$ if and only if for each $\epsilon > 0$ there exists a $\delta > 0$ such that for each $n \in \mathbb{N}$, and for each choice of

$$a \leqslant x_1 < y_1 \leqslant x_2 < y_2 \leqslant \ldots \leqslant x_n < y_n \leqslant b, \text{ with } \sum_1^n (y_i - x_i) < \delta,$$

we have

$$\sum_1^n |f(y_i) - f(x_i)| < \epsilon.$$

The reader should take note that continuity at a point is a *local* concept, and the $\delta > 0$ that works in collaboration with a given $\epsilon > 0$ may depend upon where in $[a, b]$ the point x_0 is located. Uniform continuity requires that there exist a suitable δ corresponding to ϵ, regardless of where in $[a, b]$ the points x and y are located, provided they are within δ of one another. Absolute continuity demands more, because $\delta > 0$ is required to be *independent* of *both* the location within $[a, b]$ of the $2n$ points $x_1, y_1, \ldots, x_n, y_n$ and the number $n \in \mathbb{N}$, provided only that

$$\sum_1^n |y_i - x_i| < \delta.$$

If we denote $E = \bigcup_1^n [x_i, y_i]$ in the definition of absolute continuity, then E is easily Lebesgue measurable, and the definition requires that

$$l(E) < \delta \implies \sum_1^n |f(y_i) - f(x_i)| < \epsilon.$$

Thus the absolute continuity of f is commonly denoted as $f \prec l$, which is read as f *is absolutely continuous with respect to Lebesgue measure*.

EXERCISES

7.9   Let

$$f(x) = \begin{cases} x^n \sin \frac{2\pi}{x} & \text{if } x \in (0, 1], \\ 0 & \text{if } x = 0, \end{cases}$$

where $n \in \mathbb{N}$. Prove the following conclusions.
   a) f is continuous at each point of $[0, 1]$.
   b) f is uniformly continuous on $[0, 1]$.
   c) f is *not* absolutely continuous on $[0, 1]$ if $n = 1$, but f is absolutely continuous provided $n > 1$. (Hint: Compare with Exercise 7.4.)

7.10   If f is absolutely continuous on $[a, b]$, prove that f has bounded variation on $[a, b]$. (Hint: If Δ is a partition of $[a, b]$, and if $S \subset [a, b]$ is a finite set of points, then the variation $v(\Delta \cup S) \geq v(\Delta)$.)

7.11   Show that the product of two absolutely continuous functions on a closed finite interval $[a, b]$ is absolutely continuous.

Definition 7.4.2   A monotone function f is said to be *singular* with respect to Lebesgue measure (written $f \perp l$) provided that f is *nonconstant*, yet $f'(x) = 0$ almost everywhere.

■ EXAMPLE 7.1

We will construct a *continuous*, singular function f, called the *Cantor function*, on $[0, 1]$. Each number $x \in [0, 1]$ can be expressed in a *ternary expansion*:

$$x = \sum_{0}^{\infty} \frac{a_n}{3^n} = a_0.a_1 a_2 \ldots a_n \ldots, \tag{7.9}$$

where each coefficient $a_n \in \{0, 1, 2\}$. The coefficients a_n are not unique without some further restriction. For example, if we allow infinite tails of 2s and also allow 1's, this would render ternary expansions of x in a nonunique manner, since

$$\sum_{p}^{\infty} \frac{2}{3^n} = \frac{1}{3^{p-1}}.$$

Thus, if $a_0 = 0$ and if $a_n = 2$ for all $n \geq 1$, then $x = 1$ and we could have used $a_0 = 1$ and $a_n = 0$ for all $n \geq 1$. The *Cantor set*, $\mathcal{C}$, defined in Exercise 3.11, can be described arithmetically by prohibiting the use of the ternary digit 1 but allowing infinite tails of 2s. The effect is the removal of *open* middle thirds that results in the Cantor set. Thus

$$\mathcal{C} = \left\{ x = \sum_{0}^{\infty} \frac{a_n}{3^n} \in [0, 1] \,\middle|\, a_n \in \{0, 2\} \,\forall n \right\}. \tag{7.10}$$

Notice that the relation $x \leqslant x'$ between two points of the Cantor set, $\mathcal{C}$, corresponds correctly with the same nonstrict inequality between the two corresponding sequences of ternary digits, using the lexicographic ordering. The Cantor set can be pictured as follows. Delete from $[0, 1]$ the open middle third, which is $\left(\frac{1}{3}, \frac{2}{3}\right)$. This deletion eliminates all x with ternary expansions having $a_1 = 1$ and leaves two closed intervals of length $\frac{1}{3}$ each.[73] Delete the open middle third from each of the two remaining pieces, which eliminates all x for which the ternary expansion has $a_2 = 1$. Continue an infinite sequence of such deletions of open middle thirds. The reader should note that the Cantor set is necessarily uncountable, because the same is true for the space of all infinite sequences on two symbols. The Cantor set includes the endpoints of the deleted open middle thirds from the construction process, but those endpoints comprise only a countable subset of the uncountable Cantor set. Thus $\mathcal{C}$ includes uncountably many points that are more difficult to picture mentally than the endpoints from the deleted middle thirds. The Cantor set is closed and nowhere dense.

Let each $x \in [0, 1]$ be expressed as in Equation (7.9). We will define the *Cantor function* f first *on* the Cantor set $\mathcal{C}$ by the following equation. If x is in fact expanded in a ternary manner as in Equation (7.10), without the use of

---

[73]The reader should note that the numbers $\frac{1}{3}$ and $\frac{2}{3}$ are not excluded.

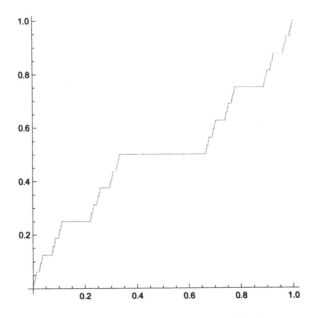

Figure 7.5   Approximation to the Cantor function.

the digit 1, then $x \in \mathcal{C}$, and we define $f(x)$ by means of the *binary expansion*

$$f(x) = \frac{1}{2} \sum_{1}^{\infty} \frac{a_n}{2^n}$$

for all $x \in \mathcal{C}$. One can see from this definition that f is monotone increasing on $\mathcal{C}$. On each of the missing open middle thirds, we define f to be locally constant. In fact, the missing open middle third (a_N, b_N) is defined by the requirement $a_N \neq 1$ on the Nth digit, and one can conclude that $f(a_N) = f(b_N)$, which is the constant value chosen for f on the closed interval $[a_N, b_N]$. For example, on the first deleted middle third f will be constantly equal to $\frac{1}{2}$, and on the next two deleted thirds f will be $\frac{1}{4}$ and $\frac{3}{4}$, respectively.

A computer rendering of the Cantor function is shown in Figure 7.5. The computer was set to connect the plotted points. This is appropriate in the sense that the Cantor function is continuous, as the reader will prove in Exercise 7.12. However, the picture is misleading as well, since it *appears* as though there were places on the graph with the derivative existing but different from zero. Actually, the derivative is zero wherever it is defined. If the picture were perfect, and if one could magnify it to an arbitrary degree, the seemingly upward-sloped parts of the graph would look just like the large-scale features, consisting of horizontal segments, except on the null set that is the Cantor set.

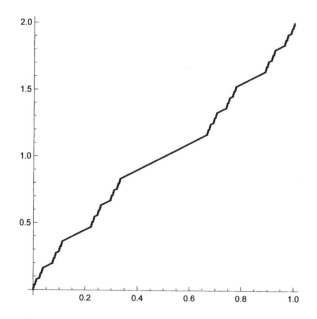

Figure 7.6   A homeomorphism that maps a measurable set to a nonmeasurable set.

EXERCISES

7.12   Show that the Cantor function f, defined in Example 7.1, maps the interval $[0, 1]$ continuously onto itself, and is a monotone increasing function for which $f'(x)$ exists and equals zero almost everywhere, and such that $f(0) = 0$ and $f(1) = 1$.

7.13   Let f be the Cantor function and define $\phi(x) = f(x) + x$ for all $x \in [0, 1]$. Let $\mathcal{C}$ denote the (middle thirds) Cantor set. (See Figure 7.6.)

    a) Prove that $\phi : [0, 1] \to [0, 2]$ is a homeomorphism. That is, prove that ϕ is injective, surjective, and *bicontinuous*.

    b) Prove that $l(\phi([0, 1] \backslash \mathcal{C})) = 1$ and that $l(\phi(\mathcal{C})) = 1$.

    c) Let P be any nonmeasurable subset of $\phi(\mathcal{C})$. (See Theorem 3.4.4 for the existence of P.) Prove that $\phi^{-1}(P)$ is a Lebesgue measurable set but not a Borel set.

7.14   Let $(X, \mathfrak{A}, \mu)$ be a complete measure space. Suppose that $\psi : X \to \mathbb{R}$ is a measurable function such that ψ^{-1} maps null sets to null sets. Prove that ψ^{-1} maps measurable sets to measurable sets.[74] Is the mapping in Exercise 7.13.c measurable?

7.15

    a) Provide an example of a function on $[0, 1]$ that is not absolutely continuous but is of bounded variation.

[74]Compare this exercise with Exercise 6.6.

b) Provide examples of two different continuous functions on $[0, 1]$ that have the same derivative *almost everywhere* and that are both equal to zero at 0.

Theorem 7.4.1 *Let f be a monotone increasing real-valued function on $[a, b]$. Then f' exists almost everywhere on $[a, b]$, and we have the following conclusions:*

 i. $f(x) - f(a) \geqslant \int_a^x f' \, dl$ *for all $x \in [a, b]$.*

 ii. *Equality holds in the inequality above if and only if f is absolutely continuous.*

Proof: The existence of f' almost everywhere follows from Theorem 7.3.1. We need to prove the two parts concerning the inequality.

 i. For almost all t,

$$f'(t) = \lim_{h \to 0} \frac{f(t + h) - f(t)}{h}.$$

Thus we can pick a sequence $h_n \to 0+$, and for each t at which $f'(t)$ exists, we have

$$f'(t) = \lim_{n \to \infty} \frac{f(t + h_n) - f(t)}{h_n}.$$

It follows that f' is equal almost everywhere to the limit of a sequence of measurable functions, which implies that f' is measurable.

Note that each difference quotient,

$$g_n(t) = \frac{f(t + h_n) - f(t)}{h_n},$$

is nonnegative, as is $f'(t)$ wherever it is exists. Next, we apply Fatou's theorem (5.4.3) as follows. For each $[c, d] \subset (a, b)$ we have

$$
\begin{aligned}
\int_c^d f'(t) \, dl(t) &= \int_c^d \lim_{n \to \infty} \frac{f(t + h_n) - f(t)}{h_n} \, dl(t) \\
&\leqslant \liminf_{n \to \infty} \int_c^d \frac{f(t + h_n) - f(t)}{h_n} \, dl(t) \\
&= \lim_{n \to \infty} \frac{1}{h_n} \left(\int_d^{d+h_n} f(t) \, dl(t) - \int_c^{c+h_n} f(t) \, dl(t) \right) \\
&= f(d) - f(c)
\end{aligned}
$$

for almost all c and d, since f is differentiable (and hence continuous) almost everywhere.[75] In the theorem, $\lim_{c \to a+} f(c) \geqslant f(a)$ because f is monotone increasing. If f is not continuous at a, then the inequality is true *a fortiori*.

 ii. Suppose first that equality holds in the inequality of part (i). The reader will prove that f is absolutely continuous in Exercise 7.16.

---

[75]If f is continuous at c and at d, then the limit indicated will be $f(d) - f(c)$.

Suppose for the opposite direction of implication that $f \prec l$, meaning that f is absolutely continuous with respect to Lebesgue measure. We need to prove that equality holds in Theorem 7.4.1. Define

$$g(x) = \int_a^x f'(t)\, dl(t),$$

so that $g \prec l$—by Exercise 7.16 again. Now let $h = f - g$, and we have

$$h'(x) = f'(x) - g'(x) = 0$$

almost everywhere. Also, it is easy to check that

$$h = f - g \prec l$$

as well. The key to the proof is to show that an absolutely continuous function with derivative equal to zero almost everywhere must be constant and, in the present case, $f(a)$. Note that because $h' = 0$ almost everywhere, the monotone function h would be singular if it were not constant. Thus we are about to show that an absolutely continuous function with zero derivative almost everywhere cannot be singular.

Since $h \prec l$, if $\epsilon > 0$ there exists $\delta > 0$ such that for *nonoverlapping* subintervals $[a_k, b_k]$ of $[a, b]$, we have

$$\sum_1^n (b_k - a_k) < \delta \implies \sum_1^n |h(b_k) - h(a_k)| < \frac{\epsilon}{2}.$$

We *claim* that h must be constant. Let

$$\mathcal{I} = \left\{ I = [c, d] \subseteq [a, b] \,\middle|\, h(I) = h(d) - h(c) < \frac{\epsilon}{2(b-a)} |I| \right\}.$$

Then $\mathcal{I}$ covers $E = \{x \in [a, b] \mid h'(x) = 0\}$ in the sense of Vitali. Hence the Vitali covering theorem (7.3.2) tells us that there exists a *disjoint* sequence $I_k \in \mathcal{I}$ such that

$$E \overset{\circ}{\subseteq} \overset{\cdot}{\bigcup_{k \in \mathbb{N}}} I_k.$$

Since $h' = 0$ almost everywhere, we see also that

$$[a, b] \overset{\circ}{\subseteq} \overset{\cdot}{\bigcup_{k \in \mathbb{N}}} I_k \subseteq [a, b],$$

which implies that $\sum_k |I_k| = b - a$. Thus

$$\sum_1^\infty h(I_k) < \frac{\epsilon}{2(b-a)} \sum_1^\infty |I_k| = \frac{\epsilon}{2}.$$

There exists $N \in \mathbb{N}$ such that $\sum_{k>N} |I_k| < \delta$. We can write

$$[a, b] \backslash \bigcup_{k \leqslant N} I_k = \bigcup_{k \leqslant p} J_k$$

for some natural number p, and where each J_k is an interval, *not closed*, and $\sum_{k \leqslant p} |\bar{J}_k| < \delta$. Because of the absolute continuity criterion for the monotone increasing function h,

$$h(b) - h(a) \leqslant \sum_{1}^{N} h(I_k) + \sum_{k \leqslant p} h(\bar{J}_k) < \frac{\epsilon}{2} + \frac{\epsilon}{2} = \epsilon$$

for all $\epsilon > 0$. Thus h must be a constant function. But then

$$f(x) - g(x) = h(x) \equiv h(a) = f(a) - g(a)$$
$$= f(a) - \int_a^a f' \, dl = f(a).$$

Thus

$$f(x) - f(a) = g(x) = \int_a^x f' \, dl.$$

∎

Corollary 7.4.1 *If f is an absolutely continuous, real-valued function on $[a, b]$, then*
$$f(x) - f(a) = \int_a^x f' \, dl \text{ for all } x \in [a, b].$$

Proof: See Exercise 7.18.a.

∎

Definition 7.4.3 A real-valued function f on a measure space $(X, \mathfrak{A}, \mu)$ is called *essentially bounded* if and only if there exists $M \in \mathbb{R}$ such that $|f(x)| \leqslant M$ for almost all x.

We denote the set of all essentially bounded functions as

$$L^{\infty}(X, \mathfrak{A}, \mu)$$

and we define the essential supremum of f by

$$\|f\|_{\infty} = \inf\{M \mid |f| \leqslant M \text{ a.e.}\}.$$

EXERCISES

7.16   Let $f \in L^1(\mathbb{R})$.
   a) If $\epsilon > 0$, prove that there exists $\delta > 0$ such that if E is Lebesgue measurable and if $l(E) < \delta$, then $\int_E |f| \, dl < \epsilon$. (Hint: Use Definition 5.2.3.)

b) Prove that if equality holds in Theorem 7.4.1, then f is absolutely continuous.

7.17 Suppose that both f and $\frac{\partial f}{\partial y}$ lie in $L^1([a, b] \times [c, d])$. Suppose also that $f(x, y)$ is absolutely continuous as a function of y for almost all fixed values of x. Prove that

$$\frac{\partial}{\partial y} \int_a^b f(x, y) \, dl(x) = \int_a^b \frac{\partial f}{\partial y}(x, y) \, dl(x)$$

for almost all y. Take care to establish that both sides exist. (Hint: Use Fubini's theorem to prove that

$$g(y) = \int_a^b f(x, y) \, dl(x) - \int_c^y \left(\int_a^b \frac{\partial f}{\partial t}(x, t) \, dl(x) \right) dl(t)$$

is a constant function of y.)

7.18 Let f be an absolutely continuous, real-valued function on $[a, b]$
  a) Prove Corollary 7.4.1. (Hint: Note that f need not be monotone. Use Equation (7.2) to express f as the difference between two *monotone* absolutely continuous functions.)
  b) Prove that the total variation of f on $[a, b]$ is *equal* to $\int_a^b |f'| \, dl$. (Hint: Use the result and the hint for Exercise 7.18.a to prove an inequality in one direction. Use Equation (7.3) to prove the opposite inequality.)

7.19 A real-valued function f on an interval I for which there exists a constant C such that

$$|f(x) - f(y)| \leqslant C|x - y|$$

for all x and y in I is called a *Lipschitz function*.
  a) Show that a Lipschitz function is absolutely continuous.
  b) Show that an absolutely continuous function f on an interval is Lipschitz if and only if f' is essentially bounded.
  c) Give an example of a Lipschitz function that does not satisfy the Mean Value Theorem for derivatives.

7.20
  a) Provide an example of a function of unbounded variation on $[0, 1]$ that has a derivative equal to zero at almost all $x \in [0, 1]$.
  b) Provide an example of a function that is absolutely continuous on $[0, 1]$ but has an unbounded derivative.

We know already that if $f \in L^1[a, b]$ and if $F(x) = \int_a^x f(t) \, dl(t)$, then $F'(x)$ exists and equals $f(x)$ almost everywhere. Thus

$$\lim_{h \to 0} \int_x^{x+h} \frac{f(t) - f(x)}{h} \, dl(t) = 0 \tag{7.11}$$

for almost all x. We have the following stronger theorem, and the reader should pause to consider why the proof is not as simple as that of Equation (7.11).

Theorem 7.4.2 (Lebesgue) *Let $f \in L^1[a, b]$. Then*

$$\lim_{h \to 0} \int_x^{x+h} \frac{|f(t) - f(x)|}{h} \, dl(t) = 0$$

for almost all x.

Proof: Suppose α is a given constant. We define the set N_α to be the minimal set such that $x \in [a, b]\backslash N_\alpha$ implies that

$$\lim_{h \to 0} \frac{1}{h} \int_x^{x+h} |f(t) - \alpha| \, dl(t) = |f(x) - \alpha|.$$

We know that N_α is a Lebesgue null set because of Theorem 7.2.1. We will show that we can choose the sets N_α independent of α. Write the set of all rational numbers as $\mathbb{Q} = \{\alpha_i \mid i \in \mathbb{N}\}$ and let $N = \bigcup_{i \in \mathbb{N}} N_{\alpha_i}$, which is a null set.

Now let β be an arbitrary real number and pick $\alpha \in \mathbb{Q}$ such that

$$|\beta - \alpha| < \epsilon.$$

We apply the triangle inequality as follows:

$$\left| \frac{1}{h} \int_x^{x+h} |f(t) - \beta| \, dl(t) - |f(x) - \beta| \right|$$

$$\leq \left| \frac{1}{h} \int_x^{x+h} |f(t) - \beta| \, dl(t) - \frac{1}{h} \int_x^{x+h} |f(t) - \alpha| \, dl(t) \right|$$

$$+ \left| \frac{1}{h} \int_x^{x+h} |f(t) - \alpha| \, dl(t) - |f(x) - \alpha| \right|$$

$$+ \left| |f(x) - \alpha| - |f(x) - \beta| \right|$$

$$< 2\epsilon + \left| \frac{1}{h} \int_x^{x+h} |f(t) - \alpha| \, dl(t) - |f(x) - \alpha| \right| \to 2\epsilon$$

as $h \to 0$ for all $x \in [a, b]\backslash N$.   ∎

Next, we consider an application.

Definition 7.4.4 Let A be a Lebesgue measurable subset of $\mathbb{R}$. A point $x \in \mathbb{R}$ is called a *density point* of A if and only if

$$\lim_{h \to 0+} \frac{l(A \cap [x - h, x + h])}{2h}$$

exists and equals 1.

A density point of A need not belong to A.

EXERCISE

7.21   Let A be a Lebesgue measurable subset of $\mathbb{R}$ of positive measure.

   a) Apply Theorem 7.4.2 to the function $f = 1_A$, the indicator function of A, in order to prove that almost every point $x \in A$ is a density point of A. [76]

   b) Suppose that A and B are two sets of strictly positive measure in $\mathbb{R}$. Apply the preceding part to prove that there exists a translation by some $h \in \mathbb{R}$ such that $l((A + h) \cap B) > 0$. [77] (Hint: Consider two density points.)

The following surprising congruence theorem is a fairly simple consequence of any one of the Exercises 7.21, 3.26, or 6.11.

Theorem 7.4.3   (Steinhaus) *Let A and B be any two subsets of $\mathbb{R}$ having identical, finite positive measure: $l(A) = l(B) = \alpha$ and $0 < \alpha < \infty$. Then there exist two sequences of mutually disjoint measurable sets A_n and B_n and null sets N and M such that*

$$A = \overset{\cdot}{\bigcup_{n \in \mathbb{N}}} A_n \cup N,$$

$$B = \overset{\cdot}{\bigcup_{n \in \mathbb{N}}} B_n \cup M,$$

and there exist constants a_n such that $A_n + a_n = B_n$ for all $n \in \mathbb{N}$.

Proof: The function

$$f(x) = l((A + x) \cap B)$$

is a continuous function of x which approaches zero as $|x| \to \infty$ and which achieves strictly positive values at least for some x. Thus there exists a number $x = a_1$ which maximizes the value of f. Let $B_1 = B \cap (A + a_1)$, and let $A_1 = B_1 - a_1$. Define $B^1 = B \backslash B_1$ and $A^1 = A \backslash A_1$. If B^1 and A^1 happen to be null sets, we are done.

   If not, pick a_2 which maximizes $l((A^1 + x) \cap B^1)$ and define A_2, B_2, A^2, and B^2 in the same manner as in the first step. We proceed until the process terminates (in which case we are done) or else we generate in this way two infinite sequences of sets and translation numbers. In the latter case, observe that

$$l(A_n) = l(B_n) \to 0$$

as $n \to \infty$. Let

$$N = A \backslash \overset{\cdot}{\bigcup_{n \in \mathbb{N}}} A_n$$

and let

$$M = B \backslash \overset{\cdot}{\bigcup_{n \in \mathbb{N}}} B_n.$$

---

[76]It is interesting to compare the result of this exercise concerning density points with Exercise 3.17.

[77]This part calls for a new proof of a theorem the reader has proven in an earlier exercise by a different method. See either Exercise 3.26 or Exercise 6.11.

It will suffice to prove that N and M, which must have the same measure, are null sets.

Suppose this conclusion were false. Then there exists $a \in \mathbb{R}$ such that

$$l((N + a) \cap M) > 0.$$

But then there exists n such that

$$l(A_n) = l(B_n) < l((N + a) \cap M).$$

This violates the maximality property in the choice of a_n.   ∎

CHAPTER 8

GENERAL COUNTABLY ADDITIVE SET FUNCTIONS

In Theorem 5.2.2 the reader saw that if $f : X \to \mathbb{R}$ is integrable on the measure space $(X, \mathfrak{A}, \mu)$, then we can define a countably additive set function ν on $\mathfrak{A}$ by the formula

$$\nu(A) = \int_A f \, d\mu. \tag{8.1}$$

We learned that the set function ν can take both positive and negative values and that ν is bounded in absolute value by $\|f\|_1$.

In this chapter we will study general countably additive set functions that can take both positive and negative values. Such set functions are known also as *signed measures*. In the *Radon-Nikodym* theorem we will characterize all those signed measures that arise from integrals of an integrable function with respect to a measure, μ, as in Equation (8.1), as being *absolutely continuous* with respect to the measure μ. And in the *Lebesgue Decomposition* theorem, we will learn how to decompose any signed measure into its *absolutely continuous* and *singular* parts, with respect to a given measure μ. These concepts for signed measures will be defined as part of the work of this chapter.

Measure and Integration: A Concise Introduction to Real Analysis. By Leonard F. Richardson
Copyright © 2009 John Wiley & Sons, Inc.

8.1 HAHN DECOMPOSITION THEOREM

Definition 8.1.1 Given a σ-algebra $\mathfrak{A}$ of subsets of X, a function $\mu : \mathfrak{A} \to \mathbb{R}$ is called a *countably additive set function* (or a *signed measure*),[78] provided that for every sequence of mutually disjoint sets $A_n \in \mathfrak{A}$, we have

$$\mu\left(\bigcup_{n \in \mathbb{N}} A_n\right) = \sum_{n \in \mathbb{N}} \mu(A_n).$$

We prove first the following theorem.

Theorem 8.1.1 *If μ is a countably additive set function on a σ-field $\mathfrak{A}$, then μ is bounded on $\mathfrak{A}$. That is, there exists a real number M such that $|\mu(A)| \leqslant M$ for all $A \in \mathfrak{A}$.*

Proof: We begin by restating the theorem as follows, bearing in mind that μ can have *both* positive and negative values on $\mathfrak{A}$, and that consequently μ *need not be monotone*. Let

$$\mu_*(A) = \sup\left\{|\mu(B)| \,\big|\, B \subset A, B \in \mathfrak{A}\right\}$$

for each $A \in \mathfrak{A}$. The theorem asserts that $\mu_*(X) < \infty$.

i. We claim that both $|\mu(A)|$ and $\mu_*(A)$ are subadditive as functions of $A \in \mathfrak{A}$.[79] The inequality for $|\mu|$ follows immediately from the triangle inequality for the real numbers combined with the additivity of μ: If A and B are $\mathfrak{A}$-measurable and disjoint, then

$$|\mu(A \,\dot\cup\, B)| = |\mu(A) + \mu(B)| \leqslant |\mu(A)| + |\mu(B)|.$$

For the second inequality, we note that if $A = A_1 \,\dot\cup\, A_2$, a disjoint union, then $\mu_*(A) \leqslant \mu_*(A_1) + \mu_*(A_2)$ because if an $\mathfrak{A}$-measurable set $B \subset A$, then

$$|\mu(B)| \leqslant |\mu(B \cap A_1)| + |\mu(B \cap A_2)|$$
$$\leqslant \mu_*(A_1) + \mu_*(A_2).$$

Here, we have used the subadditivity of $|\mu(B)|$ as a function of B.

ii. We will suppose that $\mu_*(X) = \infty$ and deduce a contradiction. By hypothesis, $\mu(X) \in \mathbb{R}$. So there exists a set $B \in \mathfrak{A}$ such that

$$|\mu(B)| > |\mu(X)| + 1 \geqslant 1.$$

---

[78] Some authors allow a signed measure to be *extended* real-valued. In that case, it is necessary to require that μ take at most *one* of the two infinite values, ∞ or $-\infty$, in order to ensure that μ is well defined on $\mathfrak{A}$. We will restrict ourselves to real-valued, countably additive set functions here, however.

[79] We do *not* denote $|\mu(A)|$ in the form $|\mu|(A)$ because the latter symbol will be given a special meaning in Definition 8.1.2.

By the additivity of μ,

$$|\mu(X\backslash B)| = |\mu(X) - \mu(B)|$$
$$\geq |\mu(B)| - |\mu(X)| > 1.$$

Because B and $X\backslash B$ are disjoint, it follows from subadditivity that either

$$\mu_*(B) = \infty \ \text{ or } \ \mu_*(X\backslash B) = \infty.$$

Thus there exists $B_1 \in \mathfrak{A}$ such that $|\mu(B_1)| > 1$ *and* $\mu_*(X\backslash B_1) = \infty$. Hence there exists $B_2 \in \mathfrak{A}$, disjoint from B_1, such that

$$|\mu(B_2)| > 1 \ \text{ and } \ \mu_*(X\backslash(B_1 \cup B_2)) = \infty.$$

This process generates an infinite sequence of mutually disjoint sets $B_n \in \mathfrak{A}$ such that $|\mu(B_n)| > 1$ for each $n \in \mathbb{N}$. Let

$$\mathcal{B} = \dot{\bigcup_{n\in\mathbb{N}}} B_n,$$

so that

$$\mu(\mathcal{B}) = \sum_{n\in\mathbb{N}} \mu(B_n). \tag{8.2}$$

The latter series is conditionally convergent, meaning that it is convergent but *not* absolutely convergent. Therefore, by a familiar exercise or theorem from advanced calculus,[80] both the sum of the positive terms and the sum of the negative terms in Equation (8.2) must diverge. Hence there exists a subsequence B_{n_j} such that

$$\mu\left(\bigcup_{j\in\mathbb{N}} B_{n_j}\right) \notin \mathbb{R},$$

which is a contradiction.

∎

We are ready to state and prove the Hahn Decomposition theorem.

Theorem 8.1.2 (Hahn Decomposition) *Let μ be a countably additive set function on a σ-algebra $\mathfrak{A}$. Then there exists a partition, $X = P\dot{\cup}N$ into disjoint measurable sets, with the following properties for each $A \in \mathfrak{A}$:*

i. *If $A \subseteq P$, then we must have $\mu(A) \geq 0$.*

ii. *If $A \subseteq N$, then $\mu(A) \leq 0$.*

---

[80]See, for example, [20]. There it is shown that if a series is conditionally convergent, then the sum of the positive terms diverges and the sum of the negative terms diverges.

iii. The partition $X = P \,\dot\cup\, N$ is essentially unique in the following sense. If $X = P' \cup N'$ is another such decomposition, then each measurable subset of $P \,\triangle\, P'$ is a μ-null set, and each measurable subset of $N \,\triangle\, N'$ is a μ-null set.

We observe that each measurable subset of $P \triangle P'$ has nonnegative measure, whereas each measurable subset of $N \triangle N'$ has nonpositive measure. Thus the essential uniqueness criterion can be restated as follows:

$$\mu(P \triangle P') = 0 = \mu(N \triangle N').$$

Proof: Let

$$\alpha = \sup\{\mu(A) \mid A \in \mathfrak{A}\}. \tag{8.3}$$

Then $0 \leqslant \alpha < \infty$ by Theorem 8.1.1 and because $\mu(\varnothing) = 0$. For each $n \in \mathbb{N}$ there exists $A_n \in \mathfrak{A}$ such that

$$\mu(A_n) > \alpha - \frac{1}{2^n}.$$

Let

$$P = \liminf A_n = \bigcup_{p=1}^{\infty} \bigcap_{n=p}^{\infty} A_n,$$

so that $P \in \mathfrak{A}$ and P is the set of all those $x \in X$ such that x is present in all but a finite number of the sets A_n. We will show that $\mu(P) = \alpha$. First, we need the following lemma.

Lemma 8.1.1 *Under the hypotheses of Theorem 8.1.2, with α defined by Equation (8.3), if $\mu(B_1) > \alpha - \epsilon_1$ and if $\mu(B_2) > \alpha - \epsilon_2$, then*

$$\mu(B_1 \cap B_2) > \alpha - (\epsilon_1 + \epsilon_2).$$

Proof: Because μ is additive,

$$
\begin{aligned}
\mu(B_1 \cap B_2) &= \mu(B_1) + \mu(B_2) - \mu(B_1 \cup B_2) \\
&> (\alpha - \epsilon_1) + (\alpha - \epsilon_2) - \alpha \\
&= \alpha - (\epsilon_1 + \epsilon_2)
\end{aligned}
$$

because of Equation (8.3). ∎

Letting

$$H_q = \bigcap_{n=p}^{p+q} A_n \quad \text{and} \quad H_{\infty} = \bigcap_{n=p}^{\infty} A_n,$$

we see that $H_q \supseteq H_{q+1}$ for all $q \in \mathbb{N}$. Also,

$$\mu(H_q) > \alpha - \sum_{n=p}^{p+q} \frac{1}{2^n}$$

$$> \alpha - \frac{1}{2^{p-1}}$$

for all $q \in \mathbb{N}$. By countable additivity of μ, we see that

$$\mu(H_\infty) = \mu(H_1) - \sum_{q=1}^{\infty} \mu(H_q \backslash H_{q+1})$$

$$= \lim_{N \to \infty} \left(\mu(H_1) - \sum_{q=1}^{N} \mu(H_q \backslash H_{q+1}) \right)$$

$$= \lim_{N \to \infty} \mu(H_{N+1}).$$

We deduce that

$$\mu \left(\bigcap_p^{\infty} A_n \right) = \lim_{N \to \infty} \mu(H_{N+1})$$

$$\geqslant \alpha - \frac{1}{2^{p-1}}.$$

It follows that

$$\mu \left(\bigcup_{p=1}^{q} \left[\bigcap_{n=p}^{\infty} A_n \right] \right) \geqslant \alpha - \frac{1}{2^{q-1}}$$

since the union over p is the union of an increasing chain of sets. Since μ is countably additive,

$$\alpha - \frac{1}{2^{q-1}} \leqslant \mu(P) \leqslant \alpha$$

for all $q \in \mathbb{N}$, which implies that $\mu(P) = \alpha$, as claimed.

Moreover, if there were a set $A \in \mathfrak{A}$ such that $A \subseteq P$ and $\mu(A) < 0$, then we would have

$$\mu(P \backslash A) = \mu(P) - \mu(A) > \mu(P),$$

which is a contradiction. It follows that if $A \subseteq P$, then $\mu(A) \geqslant 0$. Now let $N = X \backslash P$. Suppose there were a measurable set $A \subseteq N$ such that $\mu(A) > 0$. Then it would follow that $\mu(P \cup A) > \mu(P)$, which is impossible. Hence if $A \in \mathfrak{A}$ and $A \subseteq N$, it follows that $\mu(A) \leqslant 0$. We leave the proof of essential uniqueness to Exercise 8.1. ∎

Definition 8.1.2 Let μ be a countably additive set function on a σ-algebra $\mathfrak{A}$, and let P and N be (for μ) as in Theorem 8.1.2. Define the *positive part*, the *negative part*, and the *variation* of μ as follows:

$$\begin{aligned}
\mu^+(A) &= \mu(A \cap P) \\
\mu^-(A) &= |\mu(A \cap N)| \\
|\mu|(A) &= \mu^+(A) + \mu^-(A).
\end{aligned}$$

The number $|\mu|(X) = \|\mu\|$ is called the *total variation norm* of μ. See Exercise 8.3.

EXERCISES

8.1 Suppose that we have two Hahn decompositions as in Theorem 8.1.2:

$$X = P \cup N = P' \cup N'.$$

Prove that $\mu(P \bigtriangleup P') = 0 = \mu(N \bigtriangleup N')$.

8.2 Prove the *Jordan Decomposition Theorem* as follows:
   a) Prove that μ^+, μ^-, and $|\mu|$, as in Definition 8.1.2, are countably additive nonnegative measures.
   b) Prove that the decomposition

$$\mu = \mu^+ - \mu^-$$

   is *minimal* in the following sense. If μ_1 and μ_2 are measures such that $\mu = \mu_1 - \mu_2$, then $\mu^+ \leqslant \mu_1$ and $\mu^- \leqslant \mu_2$.

8.3 Prove that the total variation norm, as in Definition 8.1.2, satisfies all the requirements to be a norm on the vector space $\mathcal{M}$ of all countably additive set functions on $(X, \mathfrak{A})$, a measurable space consisting of the set X and a σ-field $\mathfrak{A}$ of subsets of X.

8.4 Prove that $\mathcal{M}$ is complete in the total variation norm.

8.2 RADON-NIKODYM THEOREM

Definition 8.2.1 If λ and μ are *measures* on a σ-algebra $\mathfrak{A}$ of subsets of X, we call λ *absolutely continuous* with respect to μ, written as $\lambda \prec \mu$, if and only if $\mu(A) = 0$ implies $\lambda(A) = 0$ for all $A \in \mathfrak{A}$.

If a nonnegative function f is in $L^1(X, \mathfrak{A}, \mu)$, and if we define

$$\mu_f(E) = \int_E f \, d\mu$$

for each $E \in \mathfrak{A}$, then μ_f will be absolutely continuous with respect to μ, written $\mu_f \prec \mu$, as in the foregoing definition.

We have a similar definition for the absolute continuity of one countably additive set function (signed measure) with respect to another.

Definition 8.2.2 If λ and μ are *countably additive set functions* on $\mathfrak{A}$, we call λ *absolutely continuous* with respect to μ, written $\lambda \prec \mu$, if and only if $\lambda(E) = 0$ for each $E \in \mathfrak{A}$ such that $|\mu|(E) = 0$.

See Exercise 8.5.

Theorem 8.2.1 *Suppose λ and μ are finite (nonnegative) measures on a σ-algebra $\mathfrak{A}$ of subsets of a set X. Then we have the following conclusions:*

i. *The measure $\lambda \prec \mu$ if and only if there exists a nonnegative function f in $L^1(X, \mathfrak{A}, \mu)$, called the* Radon-Nikodym derivative *and denoted by* $\dfrac{d\lambda}{d\mu}$, *such that for each $A \in \mathfrak{A}$ we have*

$$\lambda(A) = \int_A f \, d\mu = \int_A \frac{d\lambda}{d\mu} \, d\mu.$$

ii. *Moreover, the $L^1(X, \mathfrak{A}, \mu)$-equivalence class of a Radon-Nikodym derivative, $\dfrac{d\lambda}{d\mu}$, is uniquely determined.*

Remark 8.2.1 The notation for the Radon-Nikodym derivative suggests a chain rule (Exercise 8.9) and a change of variables formula (Exercise 8.13).

Proof: The implication from right to left in part (i) is inherent in the fourth conclusion of Theorem 5.2.2. So we will suppose here that $\lambda \prec \mu$, and we give a proof from left to right. We begin with a lemma.

Lemma 8.2.1 *Under the hypotheses of the Radon-Nikodym theorem, if*

$$\lambda \prec \mu,$$

and if the measure λ is not identically zero, then there exists $\epsilon > 0$, and there exists $P \in \mathfrak{A}$ with $\mu(P) > 0$, such that if $A \in \mathfrak{A}$ and $A \subset P$, we have $\lambda(A) \geq \epsilon\mu(A)$.

In the context of the proof of this lemma, we will write the conclusion of the lemma as an inequality as follows: $\lambda \overset{P}{\geq} \epsilon\mu$.

Proof: Since $\lambda \prec \mu$ and λ is not identically zero, neither is μ identically zero. We claim that there exists sufficiently small $\epsilon > 0$ such that the nonnegative measure

$$(\lambda - \epsilon\mu)^+ \neq 0,$$

meaning that $(\lambda - \epsilon\mu)^+$ is strictly positive on some set. Suppose this were false. Then we would conclude from the Hahn Decomposition theorem that $\lambda(A) \leq \epsilon\mu(A)$ for all $A \in \mathfrak{A}$, and for all $\epsilon > 0$. But this would force $\lambda \equiv 0$, which would be a contradiction.

Thus there exists $\epsilon > 0$ such that $(\lambda - \epsilon\mu)^+ \neq 0$. Hence there is a Hahn Decomposition $X = P \cup N$ for the signed measure $\lambda - \epsilon\mu$, with $\mu(P) > 0$, since $(\lambda - \epsilon\mu)^+(P) > 0$. If a measurable set $A \subset P$, then $(\lambda - \epsilon\mu)(A) \geq 0$. Thus there exists a set P of strictly positive μ-measure, for which

$$\lambda \overset{P}{\geq} \epsilon\mu.$$

∎

If $f \in L^+(\mu)$, we define $\mu_f(A) = \int_A f \, d\mu$, as we did earlier, and we let

$$L^+(\mu, \lambda) = \{ f \in L^+(\mu) \mid \mu_f(A) \leqslant \lambda(A) \; \forall A \in \mathfrak{A} \},$$

noting that the inequalities that define $L^+(\mu, \lambda)$ apply to all $A \in \mathfrak{A}$. By Lemma 8.2.1, we know that there exist $\epsilon > 0$ and $P \in \mathfrak{A}$, with $\mu(P) > 0$, and such that

$$\epsilon 1_P \in L^+(\mu, \lambda),$$

which therefore has a nontrivial element if λ is not identically zero. Even if λ were zero, the set $L^+(\mu, \lambda)$ would be nonempty since it would contain the zero function.

Let

$$\alpha = \sup\{\mu_f(X) \mid f \in L^+(\mu, \lambda)\},$$

so that $\alpha \leqslant \lambda(X) < \infty$. Thus, for each $n \in \mathbb{N}$, there exists $f_n \in L^+(\mu, \lambda)$ such that $\mu_{f_n}(X) > \alpha - \frac{1}{n}$. Let

$$g_n = \max(f_1, \ldots, f_n) = f_1 \vee \ldots \vee f_n \in L^+(\mu, \lambda).$$

Then g_n is a monotone increasing sequence of measurable functions. Moreover, $\mu_{g_n} \leqslant \lambda$ and $\mu_{g_n}(X) > \alpha - \frac{1}{n}$.

We can define $g = \lim_n g_n$, which is defined and finite almost everywhere,[81] and we see that $\mu_g \leqslant \lambda$. We know also that $\mu_g(X) = \alpha$. It will suffice to prove that $\mu_g = \lambda$. We will suppose that the latter equation is false and deduce a contradiction.

Suppose that $\lambda - \mu_g > 0$, which means that $\lambda - \mu_g$ is nonnegative and not the identically zero measure. Let $\lambda^* = \lambda - \mu_g$. Then $\lambda^* > 0$ and $\lambda^* < \mu$. By Lemma 8.2.1, we conclude that there exists $\epsilon' > 0$ and $P' \in \mathfrak{A}$ such that $\mu(P') > 0$, and such that $A \in \mathfrak{A}$ and $A \subset P'$ imply that

$$\lambda(A) - \mu_g(A) = \lambda^*(A) \geqslant \epsilon' \mu(A).$$

Let $h = g + \epsilon' 1_{P'}$. Then $\int_A h \, d\mu = \int_A g \, d\mu + \epsilon' \mu(A)$ for each $A \in \mathfrak{A}$ such that $A \subset P'$. That is

$$\lambda \geqslant \mu_g + \mu_{\epsilon' 1_{P'}}.$$

Hence $\int_X h \, d\mu > \alpha$, which is impossible since h must lie in $L^+(\mu, \lambda)$.

The uniqueness of the Radon-Nikodym derivative up to L^1-equivalence is shown in Exercise 8.6. ∎

Remark 8.2.2 We remark that if $f \in L^1(X, \mathfrak{A}, \mu)$ for some measure μ, then f has a σ-finite carrier.[82] Thus in order to characterize those measures expressible in the form $\lambda(A) = \int_A f \, d\mu$, it would be appropriate to limit our attention to σ-finite measure spaces $(X, \mathfrak{A}, \mu)$. It is easy to extend the Radon-Nikodym theorem to the case in which μ is a σ-finite measure and λ is a finite measure.

---

[81] Here we use the Monotone Convergence theorem, together with the finiteness of $\lambda(X)$.

[82] We can take for the carrier the set $|f|^{-1}(0, \infty]$.

It is simple also to give an extension of the Radon-Nikodym theorem to signed measures because each signed measure is the difference between two positive measures. Thus the Radon-Nikodym derivative in this more general context is the difference between two Radon-Nikodym derivatives for positive measures. See Exercises 8.7 and 8.8.

EXERCISES

8.5   Let λ and μ be countably additive set functions. Prove that the following three statements are equivalent.

    a) $\lambda \prec \mu$.

    b) $\lambda^+ \prec \mu$ and $\lambda^- \prec \mu$.

    c) $|\lambda| \prec |\mu|$.

8.6   Show that $\dfrac{d\lambda}{d\mu}$, the Radon-Nikodym derivative of λ with respect to μ in Theorem 8.2.1, is uniquely determined as an element of $L^1(X, \mathfrak{A}, \mu)$.

8.7   Suppose μ is a σ-finite *measure* on a σ-algebra $\mathfrak{A}$ of subsets of X. Suppose λ is another σ-finite measure on $\mathfrak{A}$ such that $\lambda \prec \mu$. Prove that there exists a nonnegative μ-measurable function f on X such that $\lambda(A) = \int_A f\, d\mu$ for all $A \in \mathfrak{A}$. Prove that λ is a finite measure if and only if $f \in L^1(X, \mathfrak{A}, \mu)$.

8.8   Suppose μ is a σ-finite *measure* on a σ-algebra $\mathfrak{A}$ of subsets of X. Suppose λ is a *signed* real-valued measure on $\mathfrak{A}$ such that $\lambda \prec \mu$. Prove that there exists a (signed) function $f \in L^1(X, \mathfrak{A}, \mu)$ such that $\lambda(A) = \int_A f\, d\mu$ for all $A \in \mathfrak{A}$.

8.9   Suppose that the measures λ, μ, ν on a measurable space $(X, \mathfrak{A})$ have the relationship

$$\lambda \prec \mu \prec \nu,$$

meaning that $\lambda \prec \mu$ and $\mu \prec \nu$, where λ and μ are finite and ν is σ-finite. Prove that $\lambda \prec \nu$ and that

$$\frac{d\lambda}{d\nu} = \frac{d\lambda}{d\mu}\frac{d\mu}{d\nu}.$$

This can be done by means of the steps below. (You may use the result of Exercise 8.7.)

    a) Be sure to show (easily) that $\lambda \prec \nu$. Let

$$f = \frac{d\lambda}{d\mu},\ g = \frac{d\mu}{d\nu},\ h = \frac{d\lambda}{d\nu},$$

    and explain why there is a monotone nondecreasing sequence $f_n \in \mathfrak{S}_0$ such that $f_n \to f$ pointwise everywhere.

    b) Show that

$$\left| \lambda(A) - \int_A f_n\, d\mu \right| \to 0$$

for all $A \in \mathfrak{A}$ as $n \to \infty$.

c) Show that $\int_A f_n \, d\mu = \int_A f_n \, g \, d\nu$ for all $A \in \mathfrak{A}$.

d) Use Exercise 8.6 to complete the proof that $f_n g_n \to h$.

8.10   Give another proof of Exercise 5.24 using a Radon-Nikodym derivative. That is, let $E_1 \supset E_2 \supset \ldots \supset E_n \supset \ldots$ be a decreasing nest of measurable sets in the complete measure space $(X, \mathfrak{A}, \mu)$. Let f be integrable on $(X, \mathfrak{A}, \mu)$ and suppose that

$$\mu\left(\bigcap_{n \in \mathbb{N}} E_n\right) = 0.$$

Prove that $\int_{E_n} f \, d\mu \to 0$ as $n \to \infty$ by letting ν be defined by $\dfrac{d\nu}{d\mu} = f$.

8.11   Let l denote Lebesgue measure on the unit interval and let f denote the Cantor function from Example 7.1.

a) Does it make sense to define $\lambda = l \circ f$ by $\lambda(A) = l(f(A))$ for each measurable set $A \subset [0, 1]$? Why or why not?

b) Suppose $A \subseteq [0, 1]$ is a μ-null set with the property that $f(A)$ is Lebesgue measurable. Must $\lambda(A) = 0$? Why or why not?

8.12   Let ϕ be a continuously differentiable monotone increasing function defined on $[a, b] \subset \mathbb{R}$. Define a measure λ on the Lebesgue measurable sets of $[a, b]$ by $\lambda(A) = l(\phi(A))$. Prove that $\lambda \prec l$ and find $\frac{d\lambda}{dl}$.

8.13   Suppose $(X, \mathfrak{A}, \mu)$ is a complete measure space and

$$f \in L^1(X, \mathfrak{A}, \mu).$$

Suppose $\phi : X \to X$ is a bijection for which $\phi(E) \in \mathfrak{A}$ if and only if $E \in \mathfrak{A}$. Suppose ϕ maps μ-null sets to μ-null sets. Define the measure

$$\mu \circ \phi(E) = \mu(\phi(E)).$$

Show that $\mu \circ \phi \prec \mu$, and prove the change of variables formula

$$\int_E (f \circ \phi) \frac{d(\mu \circ \phi)}{d\mu} \, d\mu = \int_{\phi(E)} f \, d\mu.$$

8.14   It is interesting to consider the relationship between the concept of absolute continuity of functions given in Definition 7.4.1 and that of absolute continuity of measures.

a) If λ and μ are any two finite measures on a σ-field $\mathfrak{A} \subset \mathfrak{P}(X)$, prove that $\lambda \prec \mu$ if and only if they satisfy the following condition: For each $\epsilon > 0$ there exists a $\delta > 0$ such that $\mu(A) < \delta$ implies that $\lambda(A) < \epsilon$. (Hint: For one direction, use the Radon-Nikodym theorem.)

b) Suppose now that the finite measure λ is defined on the Lebesgue measurable sets of $([a, b), \mathfrak{L})$. Define $f(x) = \lambda[a, x)$ for all $x \in [a, b]$. Prove that

f is an absolutely continuous function on $[a, b]$ if and only if $\lambda \prec l$. (Hint: From right to left is easy by part (a). For the other direction, prove that $\lambda(A) = \int_A f' \, dl$ for all $A \in \mathcal{L}$.)

8.3 LEBESGUE DECOMPOSITION THEOREM

The Radon-Nikodym theorem addressed the classification of measures absolutely continuous with respect to a given measure. Here we study a quite different (symmetrical) relationship of singularity between two measures or countably additive set functions.

Definition 8.3.1 If λ and μ are *countably additive set functions* on a measurable space $(X, \mathfrak{A})$, we call λ *singular* with respect to μ if and only if $X = E \dot\cup F$, a disjoint union of $\mathfrak{A}$-measurable sets, such that $|\lambda|(E) = 0 = |\mu|(F)$. This is denoted as

$$\lambda \perp \mu.$$

Theorem 8.3.1 (Lebesgue Decomposition Theorem) *Let $\mathfrak{A}$ be a σ-algebra of subsets of X. Let μ and ν be two signed measures defined on $\mathfrak{A}$. Then there exist two* unique *signed measures, ν_s and ν_a, such that*

$$\nu = \nu_s + \nu_a,$$

with the properties that $\nu_s \perp \mu$ and $\nu_a \prec \mu$.

Proof: It follows from Definitions 8.2.2 and 8.3.1 that singularity or absolute continuity with respect to a signed measure μ means singularity or absolute continuity with respect to $|\mu|$. Thus we can assume without loss of generality that μ is a (nonnegative) measure. Because of Exercise 8.5 and Definition 8.3.1, we can assume without loss of generality that ν is a measure as well.[83]

The proof of the theorem is based upon the simple observation that $\nu \prec (\mu + \nu)$. Thus there exists a nonnegative measurable function f such that

$$\nu(E) = \int_C f \, d(\mu + \nu) = \int_E f \, d\mu + \int_E f \, d\nu$$

for all $E \in \mathfrak{A}$. Since μ and ν are positive, we have

$$0 \leqslant \nu(E) \leqslant \mu(E) + \nu(E),$$

---

[83] By Definition 8.1.1, our assumption implies that μ and ν are finite measures. For measures that are not signed, the present theorem can be generalized readily to the σ-finite case. See Exercise 8.18.

which implies that we have $0 \leqslant f \leqslant 1$ almost everywhere with respect to $\mu + \nu$. Let $A = f^{-1}\{1\}$, and let $B = X \backslash A$. Observe that A and B are complementary, and $0 \leqslant f < 1$ almost everywhere on B. Thus

$$\nu(A) = \mu(A) + \nu(A),$$

which implies that $\mu(A) = 0$. Define

$$\nu_s(E) = \nu(E \cap A), \quad \text{and} \quad \nu_a(E) = \nu(E \cap B).$$

Thus $\nu = \nu_s + \nu_a$, and $\nu_s \perp \mu$ because ν_s vanishes on subsets of B and because $\mu(A) = 0$. We need to show that $\nu_a \prec \mu$.

Suppose that $\mu(E) = 0$. Then

$$
\begin{aligned}
\nu_a(E) = \nu(E \cap B) &= \int_{E \cap B} 1 \, d\nu \\
&= \int_{E \cap B} f \, d(\mu + \nu) \\
&= \int_{E \cap B} f \, d\nu,
\end{aligned}
$$

since $\mu(E) = 0$ by hypothesis. This implies that

$$\int_{E \cap B} (1 - f) \, d\nu = 0.$$

Since $1 - f > 0$ ν-almost everywhere on B, it follows that $\nu_a(E) = \nu(E \cap B) = 0$, so that $\nu_a \prec \mu$.

Finally, we prove uniqueness. Let $\nu = \nu_s + \nu_a = \bar{\nu}_s + \bar{\nu}_a$ be two Lebesgue decompositions. Thus $\nu_s - \bar{\nu}_s = \bar{\nu}_a - \nu_a$, with one side singular and the other side absolutely continuous with respect to μ. (See Exercise 8.15.) This forces both sides to be zero, which completes the proof. (See Exercise 8.16.) ∎

EXERCISES

8.15 Prove that a linear combination of two measures that are absolutely continuous with respect to μ on $(X, \mathfrak{A})$ must be absolutely continuous. Prove also that a linear combination of two measures that are singular with respect to μ on $(X, \mathfrak{A})$ must be singular with respect to μ.

8.16 Let μ and ν be nonnegative finite measures on $(X, \mathfrak{A})$. If $\nu \perp \mu$ and $\nu \prec \mu$, prove that $\nu = 0$, the identically zero measure on $\mathfrak{A}$.

8.17 This exercise continues the work begun in Exercise 8.14. Let f be a monotone increasing function on $[a, b]$ and define a measure μ by letting it assign to an interval $[a, x]$ the measure $\mu[a, x] = f(x) - f(a)$.

   a) Let μ_a be the absolutely continuous part of μ with respect to Lebesgue measure, and find the Radon-Nikodym derivative

$$\frac{d\mu_a}{dl}.$$

b) Show that the singular part μ_s and the absolutely continuous part μ_a of μ_f can be used to define absolutely continuous and singular parts of the function f.

8.18 Let ν be any σ-finite measure on the measure space $(X, \mathfrak{A}, \mu)$, where μ is σ-finite. Prove that there exist two *unique* measures ν_s and ν_a such that

$$\nu = \nu_s + \nu_a$$

with the properties that $\nu_s \perp \mu$ and $\nu_a < \mu$.

CHAPTER 9

EXAMPLES OF DUAL SPACES FROM MEASURE THEORY

We have seen that, for any measure space $(X, \mathfrak{A}, \mu)$, the space $L^1(X, \mathfrak{A}, \mu)$ is a Banach space. In this chapter, we will define an infinite family of Banach spaces. We will see that some of these spaces can be used to represent the Banach space of all the continuous linear mappings of other such spaces into the field of scalars.

9.1 THE BANACH SPACE $L^p(X, \mathfrak{A}, \mu)$

Recall that for a measurable function f on a measure space $(X, \mathfrak{A}, \mu)$, we define the equivalence class $[f]$ to be the set of all measurable functions g such that $f = g$ almost everywhere. This equivalence is denoted also by $f \sim g$.

Definition 9.1.1 Let $(X, \mathfrak{A}, \mu)$ be any measure space. For each real number p in the interval $[1, \infty)$, we define the *vector space*

$$L^p(X, \mathfrak{A}, \mu) = \left\{ [f] \,\middle|\, \int_X |f|^p \, d\mu < \infty, \ f \ \mathfrak{A}-\text{measurable} \right\}.$$

Measure and Integration: A Concise Introduction to Real Analysis. By Leonard F. Richardson **165**
Copyright © 2009 John Wiley & Sons, Inc.

It is common to say or to write that some *function* f is an element of $L^p(X, \mathfrak{A}, \mu)$, although the elements of that vector space are actually *equivalence classes* of functions. In Exercise 9.1, the reader will show that L^p is a vector space. It is important to define a suitable norm on $L^p(X, \mathfrak{A}, \mu)$.

Definition 9.1.2 For each f in $L^p(X, \mathfrak{A}, \mu)$, $1 \leqslant p < \infty$, we define

$$\|f\|_p = \left(\int_X |f|^p \, d\mu \right)^{\frac{1}{p}}.$$

Observe that $f \sim g$ in $L^p(X, \mathfrak{A}, \mu)$ if and only if $\|f - g\|_p = 0$. Also, $\|f\|_p$ is well defined on equivalence classes in L^p.

One of our objectives is to prove that $\| \cdot \|_p$ is a norm on the vector space L^p. The triangle inequality is the only property required of a norm that is not very easy to check. To prove the triangle inequality, we begin with an important inequality for the real numbers.

Lemma 9.1.1 (Jensen's Inequality for Real Numbers) *Suppose that*

$$\alpha \geqslant 0, \ \beta \geqslant 0, \ a > 0, \ \text{and } b > 0$$

are real numbers such that $\alpha + \beta = 1$. *Then*

$$a^\alpha b^\beta \leqslant \alpha a + \beta b.$$

Proof: We will prove this inequality from the more general measure theoretic Jensen's inequality that was proven in Theorem 5.2.3. We observe first that the logarithm function is a concave function on the positive half of the real line. We define a very simple probability space by letting $X = \{0, 1\}$ and $\mathfrak{A} = \mathfrak{P}(X)$. We define a probability measure μ such that $\mu\{0\} = \alpha$ and $\mu\{1\} = \beta$. We define an integrable function f by letting $f(0) = a$ and $f(1) = b$. It follows that

$$\log \left(\int_X f \, d\mu \right) \geqslant \int_X \log f \, d\mu.$$

Thus we see that

$$\log(\alpha a + \beta b) \geqslant \alpha \log a + \beta \log b,$$

and this implies the desired conclusion. ∎

Jensen's inequality enables us to prove the following very important inequality.

Theorem 9.1.1 (Hölder's Inequality) *Let* $p \geqslant 1$ *and* $q \geqslant 1$ *be real numbers such that*

$$\frac{1}{p} + \frac{1}{q} = 1.$$

Let $f \in L^p(X, \mathfrak{A}, \mu)$ *and* $g \in L^q(X, \mathfrak{A}, \mu)$. *Then the product* $fg \in L^1(X, \mathfrak{A}, \mu)$ *and*

$$\|fg\|_1 \leqslant \|f\|_p \|g\|_q. \tag{9.1}$$

Remark 9.1.1 In the special case in which $p = 1$, we take $q = \infty$, and we denote by $\|f\|_\infty$ the essential supremum of f. Thus $\|f\|_\infty$ is the *essential sup-norm* of f. If $p > 1$ and $q > 1$, it is common to write Hölder's inequality as

$$\int_X |fg|\, d\mu \leqslant \left(\int_X |f|^p\, d\mu \right)^{\frac{1}{p}} \left(\int_X |f|^q\, d\mu \right)^{\frac{1}{q}}.$$

Proof: Note first that if $p = 1$ and $q = \infty$, then Hölder's inequality, expressed by Equation (9.1), is very easy to prove from the monotonicity of the integral. So we will suppose that $p > 1$ and $q > 1$.

To prove Hölder's inequality, let

$$E = \left\{ x \,\middle|\, |f(x)g(x)| > 0 \right\},$$

and assume without loss of generality that $\mu(E) > 0$. We will need to consider the set E in order to have the strictly positive terms to which we can apply Jensen's inequality. For each x, let

$$a(x) = \frac{|f(x)|^p}{\|f\|_p^p} \quad \text{and} \quad b(x) = \frac{|g(x)|^q}{\|g\|_q^q},$$

where neither denominator can vanish because $\mu(E) > 0$. By Jensen's inequality for each $x \in E$, we have

$$a(x)^{\frac{1}{p}} b(x)^{\frac{1}{q}} \leqslant \frac{1}{p} a(x) + \frac{1}{q} b(x)$$

or

$$\frac{|f(x)g(x)|}{\|f\|_p \|g\|_q} \leqslant \frac{|f(x)|^p}{p\|f\|_p^p} + \frac{|g(x)|^q}{q\|g\|_q^q}.$$

Next, we integrate both sides over E to obtain

$$\frac{\|fg\|_{L^1(E)}}{\|f\|_p \|g\|_q} \leqslant \frac{1}{p} + \frac{1}{q} = 1,$$

which implies that

$$\|fg\|_1 = \|fg\|_{L^1(E)} \leqslant \|f\|_p \|g\|_q,$$

since $\|fg\|_{L^1(E)} = \|fg\|_1$, whereas the integrals of $|f|^p$ and of $|g|^q$ over E will be less than or equal to the corresponding integrals over X.   ∎

We are ready to prove the triangle inequality for the L^p-norms.

Theorem 9.1.2 (Minkowski's Inequality) *If $1 \leqslant p \leqslant \infty$ and if f and g are in $L^p(X, \mathfrak{A}, \mu)$, then*

$$\|f + g\|_p \leqslant \|f\|_p + \|g\|_p.$$

Proof: For $p = 1$ we have proven this already in Exercise 5.36. For $p = \infty$ the result is very easy. So we will suppose that $1 < p < \infty$. Because L^p is a vector space, we know that

$$0 \leqslant \|f + g\|_p < \infty.$$

If $\|f + g\|_p = 0$, then the theorem is very easy. We will suppose that $\|f + g\|_p > 0$.
    We can write

$$\int_X |f + g|^p \, d\mu = \int_X |f + g| \cdot |f + g|^{p-1} \, d\mu$$
$$\leqslant \int_X |f| \cdot |f + g|^{p-1} \, d\mu + \int_X |g| \cdot |f + g|^{p-1} \, d\mu.$$

If we let

$$q = \frac{p}{p-1}, \quad \text{then} \quad \frac{1}{p} + \frac{1}{q} = 1,$$

and we have $|f + g|^{p-1} \in L^q$. We apply Hölder's inequality to each summand on the right side above, obtaining

$$\|f + g\|_p^p \leqslant (\|f\|_p + \|g\|_p) \, \left\| |f + g|^{p-1} \right\|_q . \tag{9.2}$$

Since

$$\left\| |f + g|^{p-1} \right\|_q = \left(\int |f + g|^p \right)^{\frac{1}{q}}$$
$$= \|f + g\|_p^{p-1} \neq 0,$$

we can divide both sides of Equation 9.2 by $\|f + g\|_p^{p-1}$ to obtain Minkowski's inequality. ∎

If we introduce a metric $d(f, g) = \|f - g\|_p$, we see that L^p would be only a semimetric space if we did not employ the equivalence relation indicated in the definition. With this quotient space we have made L^p into a normed vector space. It remains to prove that this space is complete and is therefore a Banach space.

Theorem 9.1.3 *The normed vector space $L^p(X, \mathfrak{A}, \mu)$ is a Banach space for each real number $p \geqslant 1$.*

Proof: We will model the proof on that for Theorem 5.5.2. Let f_n be a Cauchy sequence in the L^p-norm. For each $k \in \mathbb{N}$, there exists a natural number n_k such that for all n and m greater than or equal to n_k we have

$$\|f_n - f_m\|_p < \frac{1}{4^{\frac{k}{p}}}.$$

In particular,

$$\|f_{n_k} - f_{n_{k+1}}\|_p < \frac{1}{4^{\frac{k}{p}}}.$$

Let

$$A_k = \left\{ x \,\middle|\, |f_{n_k}(x) - f_{n_{k+1}}(x)| \geq \frac{1}{2^{\frac{k}{p}}} \right\}.$$

Then

$$\frac{1}{2^k} \mu(A_k) \leq \int |f_{n_k} - f_{n_{k+1}}|^p \, d\mu < \frac{1}{4^k},$$

which implies that

$$\mu(A_k) < \frac{1}{2^k}.$$

Let

$$N = \limsup_k A_k = \bigcap_{m=1}^{\infty} \bigcup_{k=m}^{\infty} A_k,$$

which is the set of all those points x that appear in infinitely many sets A_k. It is easy to calculate that $\mu(N) = 0$. For all $x \notin N$, the sequence $f_{n_k}(x) \to f(x)$ and f is measurable, being the limit almost everywhere of a sequence of measurable functions. We need to prove that $f_n \to f$ in the L^p-norm.

By Fatou's lemma we have

$$\int_X |f_{n_k} - f|^p \, d\mu = \int_X \lim_j |f_{n_k} - f_{n_j}|^p \, d\mu$$

$$\leq \liminf_j \int_X |f_{n_k} - f_{n_j}|^p \, d\mu$$

$$\leq \frac{1}{4^k}.$$

Thus $f_{n_k} - f \in L^p$, which implies that $f \in L^p$. Also,

$$\|f_n - f\|_p \leq \|f_n - f_{n_k}\|_p + \|f_{n_k} - f\|_p,$$

which can be made as small as we like by choosing k sufficiently big and $n \geq n_k$. Thus $\|f_n - f\|_p \to 0$ as $n \to \infty$. ∎

EXERCISES

9.1 For each $1 \leq p < \infty$, prove that $L^p(X, \mathfrak{A}, \mu)$ is a vector space. (Hint: Closure under scalar multiplication is easy. Show that if f and g are in L^p, then $f + g \in L^p$ too. Express X as a disjoint union of two sets, depending on which of the two functions, $|f|$ or $|g|$, is larger.)

9.2 Let $(X, \mathfrak{A}, \mu)$ be measure space, and suppose that $1 \leq p \leq q \leq \infty$.
  a) Show that $L^q(X, \mathfrak{A}, \mu) \subseteq L^p(X, \mathfrak{A}, \mu)$ if $\mu(X) < \infty$.
  b) Show that if $f \in L^p(X, \mathfrak{A}, \mu)$ and if $\mu(X) = 1$, then

$$\|f\|_p \leq \|f\|_q.$$

(See Exercise 5.13.)

9.3   Let f be measurable on a measure space $(X, \mathfrak{A}, \mu)$.

  a) If $0 < \mu(X) < \infty$, prove that

$$\|f\|_p \to \|f\|_\infty$$

  as $p \to \infty$.[84]

  b) If $\mu(X) = \infty$, prove the same conclusion with the additional hypothesis that $f \in L^p$ for some $1 \leqslant p < \infty$.

9.4   Let $n \in \mathbb{N}$.

  a) Show that $L^q(\mathbb{R}^n) \nsubseteq L^p(\mathbb{R}^n)$, if $1 \leqslant p < q \leqslant \infty$.

  b) Show that $L^p(\mathbb{R}^n) \nsubseteq L^q(\mathbb{R}^n)$, if $1 \leqslant p < q \leqslant \infty$.

9.2   THE DUAL OF A BANACH SPACE

Let B be a vector space over a field $\mathbb{F}$, which may be either $\mathbb{R}$ or $\mathbb{C}$. Suppose that B is equipped with a norm, as in Definition 5.5.1. We will assume that B is also a Banach space, as in Definition 5.5.6. We have seen that for each real number $p \in [1, \infty)$ and for each measure space $(X, \mathfrak{A}, \mu)$, the space $L^p(X, \mathfrak{A}, \mu)$ is a Banach space.

Definition 9.2.1 If B is a Banach space equipped with a *norm* $\|\cdot\|$, we call $T : B \to \mathbb{F}$ a *linear functional* provided that

$$T(\alpha\mathbf{x} + \mathbf{y}) = \alpha T(\mathbf{x}) + T(\mathbf{y})$$

for all $\mathbf{x}$ and $\mathbf{y}$ in B and for all $\alpha \in \mathbb{F}$. A linear functional T is called *continuous* at $\mathbf{x}$ if and only if, for each sequence $\mathbf{x}_n$ in B such that $\mathbf{x}_n \to \mathbf{x}$, we have $T(\mathbf{x}_n) \to T(\mathbf{x})$. T is called *continuous* if and only if T is continuous at each $\mathbf{x} \in B$.

■ **EXAMPLE 9.1**

Let $\mathbf{x} = (x_1, x_2, \ldots, x_n) \in \mathbb{E}^n$, the vector space $\mathbb{R}^n$ equipped with the Euclidean norm,

$$\|\mathbf{x}\| = \sqrt{\sum_{j=1}^n x_j^2}.$$

Let $T_j : \mathbb{E}^n \to \mathbb{R}$ by the definition $T_j(\mathbf{x}) = x_j$. Then each T_j is a continuous linear functional on $\mathbb{E}^n$.

We see that the values of the continuous linear functionals T_j at $\mathbf{x}$, with j varying from 1 to n, determine $\mathbf{x}$ uniquely. In any n-dimensional Banach space, we could fix a basis arbitrarily and use $T_1(\mathbf{x}), \ldots, T_n(\mathbf{x})$ to determine the vector $\mathbf{x}$. Such a basis is not canonical, however. A canonical version of this statement could be made as follows.

[84]This exercise explains why the essential supremum of f is denoted by the symbol $\|f\|_\infty$.

Let B^* denote the vector space of all continuous linear functionals on B. Then two vectors $\mathbf{x}$ and $\mathbf{y}$ in B are distinct if and only if there exists $T \in B'$ such that $T(\mathbf{x}) \neq T(\mathbf{y})$. This can be expressed in words by saying that there are enough continuous linear functionals on a Banach space B to *separate points*. The latter statements are a consequence of the Hahn-Banach theorem, which is a very important theorem presented in many books about functional analysis. See, for example, [9]. See also Exercise 9.13.

Lemma 9.2.1 *Let B be any Banach space over $\mathbb{F}$, which may be either $\mathbb{R}$ or $\mathbb{C}$, equipped with a norm $\| \cdot \|$. A linear functional $T : B \rightarrow \mathbb{F}$ is continuous if and only if T is continuous at $\mathbf{0}$.*

Proof: If T is continuous (at all $\mathbf{x} \in B$), then it must be continuous at $\mathbf{0}$. So we prove the opposite implication. Note that since

$$T(\mathbf{0}) = T(\mathbf{0} + \mathbf{0}) = T(\mathbf{0}) + T(\mathbf{0}),$$

we must have $T(\mathbf{0}) = 0$. Suppose T is continuous at $\mathbf{0}$: That is,

$$\|\mathbf{x}_n - \mathbf{0}\| = \|\mathbf{x}_n\| \rightarrow 0$$

implies that $T(\mathbf{x}_n) \rightarrow 0 = T(\mathbf{0})$. Let $\mathbf{x} \in B$ be arbitrary and suppose that $\mathbf{x}_n \rightarrow \mathbf{x}$: That is, $\|\mathbf{x}_n - \mathbf{x}\| \rightarrow 0$. By hypothesis

$$T(\mathbf{x}_n - \mathbf{x}) = T(\mathbf{x}_n) - T(\mathbf{x}) \rightarrow 0,$$

so $T(\mathbf{x}_n) \rightarrow T(\mathbf{x})$. ∎

Definition 9.2.2 A linear functional T on a Banach space B is called *bounded* if and only if there exists a positive number $K \in \mathbb{R}$ such that $|T(\mathbf{x})| \leq K\|\mathbf{x}\|$ for all $x \in B$.[85]

It should be stressed that the same constant $K < \infty$ must suffice for all $\mathbf{x} \in B$, for T to be bounded. The reader should note that the concept of *boundedness* for a linear functional does not have the same meaning as the concept of boundedness for a function.

Theorem 9.2.1 *If T is a linear functional on a Banach space B, then T is continuous if and only if T is bounded.*

Proof: In the direction from right to left, suppose that T is bounded. It will suffice to prove that T is continuous at $\mathbf{0}$. So suppose that $\|\mathbf{x}_n\| \rightarrow 0$. Then

$$|T(\mathbf{x}_n)| \leq K\|\mathbf{x}_n\| \rightarrow 0 = T(\mathbf{0}).$$

[85]Definition 9.2.2 and Theorems 9.2.1 and 9.2.2 remain valid for any normed vector space, B, whether complete or not.

Now suppose that T is continuous. We will prove that T is bounded by contradiction. So suppose that the claim were false. Then for all $n \in \mathbb{N}$ there exists $\mathbf{x}_n \in B$ such that

$$|T(\mathbf{x}_n)\| > n\|\mathbf{x}_n\|.$$

Let

$$\mathbf{y}_n = \frac{\mathbf{x}_n}{\|\mathbf{x}_n\|\sqrt{n}},$$

where we note that $\mathbf{x}_n \neq \mathbf{0}$. Note also that

$$\|\mathbf{y}_n\| = \frac{1}{\sqrt{n}} \to 0,$$

yet $T(\mathbf{y}_n)$ fails to converge to 0. In fact,

$$|T(\mathbf{y}_n)| > \sqrt{n},$$

which is an unbounded sequence. This is a contradiction.   ∎

Because of this theorem, continuous linear functionals on normed linear spaces are often called *bounded linear functionals*.

Definition 9.2.3 Let B denote any (real or complex) Banach space. Let B' be the set of all $T : B \to \mathbb{F}$ such that T is linear and bounded. We call B' the *dual space* of B. If $T \in B'$, it has a norm defined by

$$\|T\| = \inf\left\{K \,\middle|\, |T(\mathbf{v})| \leqslant K\|\mathbf{v}\| \;\forall\, \mathbf{v} \in B\right\}.$$

See Theorem 9.2.2.

EXERCISES

9.5   Let $f \in L^1(X, \mathfrak{A}, \mu)$ and define $T(f) = \int_X f \, d\mu$. Prove that T is a bounded linear functional. Show that there is a smallest bound K and find it.

9.6   Let V be a *finite* dimensional real or complex normed vector space. Prove that every linear functional $T : B \to \mathbb{F}$ must be bounded.

9.7   Let $\mathcal{P}$ be the vector space of all *polynomials* on the interval $[0, 1]$. Give an example of a linear functional $T : \mathcal{P} \to \mathbb{R}$ such that T is not bounded with respect to the sup-norm.

9.8   Let $T \in B'$. Show that $|T(\mathbf{v})| \leqslant \|T\| \, \|\mathbf{v}\|$ for all $\mathbf{v} \in B$.

Theorem 9.2.2 *If B is any Banach space, then its dual space, B', is a Banach space.*

Proof: The reader will recall from a course in linear algebra that the sum of any two linear maps is linear, and that any constant times a linear map is linear, so we will see that B' is a vector space if we can show that the function $\| \cdot \|$ defined on B' is a norm, which will prove also that the sum of two bounded linear functionals is again bounded, and the same for scalar multiples. Observe first that $|T(\mathbf{v})| \leqslant \|T\| \cdot \|\mathbf{v}\|$ for all $\mathbf{v} \in B$, so that

$$|cT(\mathbf{v})| = |c| |T(\mathbf{v})| \leqslant |c| \cdot \|T\| \cdot \|\mathbf{v}\|.$$

This implies that

$$\|cT\| \leqslant |c| \cdot \|T\| < \infty.$$

But $T = \frac{1}{c}(cT)$ for all $c \neq 0$. Thus

$$\|T\| \leqslant \frac{1}{|c|} \|cT\|.$$

Hence $\|cT\| = |c| \|T\|$. Observe next that

$$|(T_1 + T_2)(\mathbf{v})| \leqslant |T_1(\mathbf{v})| + |T_2(\mathbf{v})| \leqslant (\|T_1\| + \|T_2\|) \|\mathbf{v}\|.$$

Thus we see that

$$\|T_1 + T_2\| \leqslant \|T_1\| + \|T_2\|.$$

To complete the proof that $\| \cdot \|$ is a norm on B', the reader should show that $\|T\| \geqslant 0$ for all T and that $\|T\| = 0$ if and only if $T = 0$.

It remains to be shown that B' is complete in the given norm. Let T_n be any Cauchy sequence in B'. Let $\epsilon > 0$. Then there exists N such that if m and n are greater than or equal to N, then $\|T_m - T_n\| < \frac{\epsilon}{2}$. Thus, for all $\mathbf{v} \in B$,

$$|T_m(\mathbf{v}) - T_n(\mathbf{v})| < \frac{\epsilon}{2} \|\mathbf{v}\|.$$

Hence $\{T_n(\mathbf{v})\}_{n=1}^{\infty}$ is a Cauchy sequence in $\mathbb{F}$ and we can define

$$T(\mathbf{v}) = \lim_{n \to \infty} T_n(\mathbf{v})$$

for all $\mathbf{v} \in B$. The proof that T is linear is an informal exercise for the student. Finally, we must show that T is bounded and that $\|T_n - T\| \to 0$. But if m and n are greater than or equal to N, we know that

$$|T_m(\mathbf{v}) - T_n(\mathbf{v})| \leqslant \frac{\epsilon}{2} \|\mathbf{v}\|.$$

Letting $n \to \infty$, we see that $|T_m(\mathbf{v}) - T(\mathbf{v})| \leqslant \frac{\epsilon}{2} \|\mathbf{v}\|$ for all $\mathbf{v} \in B$. Thus $\|T_m - T\| \leqslant \frac{\epsilon}{2} < \epsilon$, so $T_m \to T$ in the norm for B'. Moreover,

$$\|T\| = \|T_m - (T_m - T)\| \leqslant \|T_m\| + \|T_m - T\| < \infty,$$

so T is bounded as claimed. Thus B' is a Banach space. ∎

Definition 9.2.4 If f is a measurable function on the measure space $(X, \mathfrak{A}, \mu)$, define the *essential supremum* of f, denoted $\|f\|_{\infty}$, by

$$\|f\|_{\infty} = \inf \left\{ K \in \mathbb{R} \cup \{\infty\} \, \big| \, |f(x)| \leqslant K \text{ a.e.} \right\}.$$

Define $L^{\infty}(X, \mathfrak{A}, \mu)$ to be the set of all measurable functions with finite essential supremum.

EXERCISES

9.9   Let $(X, \mathfrak{A}, \mu)$ be a measure space, and suppose that $1 < p < \infty$ and that

$$\frac{1}{p} + \frac{1}{q} = 1.$$

Let $g \in L^q(X, \mathfrak{A}, \mu)$, and define $T_g(f) = \int_X fg\, d\mu$ for each f in $L^p(X, \mathfrak{A}, \mu)$.
   a) Prove that T_g is a bounded linear functional.
   b) Prove that $\|T_g\| \leqslant \|g\|_q$.

9.10   Let (X, A, μ) be a measure space.
   a) Prove that $L^{\infty}(X, \mathfrak{A}, \mu)$ is a Banach space.
   b) Prove that for each $g \in L^{\infty}(X, \mathfrak{A}, \mu)$, the function

$$T_g(f) = \int_X fg\, d\mu$$

   is a bounded linear functional on the Banach space $L^1(X, \mathfrak{A}, \mu)$.
   c) Prove that $\|T_g\| \leqslant \|g\|_{\infty}$.

9.3   THE DUAL SPACE OF $L^p(X, \mathfrak{A}, \mu)$

In Exercises 9.9 and 9.10 we saw that for each $g \in L^q(X, \mathfrak{A}, \mu)$ there is a corresponding bounded linear functional

$$T_g : f \to \int_X fg\, d\mu$$

acting on $L^p(X, \mathfrak{A}, \mu)$, provided[86] that $1 \leqslant p < \infty, 1 < q \leqslant \infty$, and that

$$\frac{1}{p} + \frac{1}{q} = 1.$$

In this section we will prove that all bounded linear functionals on L^p arise in this way, thereby characterizing the dual space of L^p in terms of L^q.

Theorem 9.3.1 *Let $(X, \mathfrak{A}, \mu)$ be any σ-finite measure space. Let $1 \leqslant p < \infty$, and suppose that*

$$\frac{1}{p} + \frac{1}{q} = 1.$$

Let T be any bounded linear functional on the Banach space $L^p(X, \mathfrak{A}, \mu)$. Then there exists a unique function g in $L^q(X, \mathfrak{A}, \mu)$ such that

$$T(f) = \int_X fg\, d\mu = T_g(f),$$

---

[86] The equation that follows is interpreted informally when $p = 1$ and $q = \infty$.

for all $f \in L^p(X, \mathfrak{A}, \mu)$. Moreover

$$\|T_g\| = \|g\|_q,$$

and the map $g \to T_g$ is a Banach space isomorphism *of L^q onto the dual space of L^p.*

Remark 9.3.1 This theorem can be described as a *representation theorem*, because it represents the dual space, $L^p(X)'$, as being *isomorphic* to $L^q(X)$. The mapping $g \to T_g$ is the *isomorphism* in the direction from L^q to $(L^p)'$. It is clear that the mapping $g \to T_g$ is linear, so that this mapping will be a Banach space isomorphism if it is onto and norm-preserving (and therefore bijective).

Proof:

1. Suppose that $\mu(X) < \infty$ and that $p > 1$, so that $q < \infty$. We will begin by obtaining the required function g for the case of a bounded *real-valued* $\mathbb{R}$-linear functional acting on *real-valued L^p-functions f*. For each set $E \in \mathfrak{A}$ define

$$\lambda(E) = T(1_E).$$

Note that $1_E \in L^p$ because $\mu(X) < \infty$, so that 1_E does lie in the domain of definition of T. We will show that λ is a countably additive set function (signed measure) on $\mathfrak{A}$. Let

$$E = \dot{\bigcup}_{k \in \mathbb{N}} E_k$$

be a disjoint union of sets in $\mathfrak{A}$. Let

$$A_n = \bigcup_1^n E_k.$$

Since

$$\sum_k \mu(E_k) = \mu(E) < \infty,$$

we know that $\mu(E \backslash A_n) \to 0$ as $n \to \infty$. It follows that

$$\|1_{A_n} - 1_E\|_{L^p(X, \mathfrak{A}, \mu)} \to 0$$

as $n \to \infty$. Since T is bounded and thus continuous, we have

$$\begin{aligned}
\lambda(E) &= T(1_E) \\
&= \lim_n T(1_{A_n}) \\
&= \lim_n \lambda(A_n) \\
&= \sum_1^\infty \lambda(E_k),
\end{aligned}$$

proving that λ is countably additive on $\mathfrak{A}$.

Moreover, if $\mu(E) = 0$, then $\|1_E\|_p = 0$, forcing $T(1_E) = \lambda(E) = 0$. Thus λ is absolutely continuous with respect to μ. By the Radon-Nikodym Theorem (8.2.1), there exists a function $g \in L^1(X, \mathfrak{A}, \mu)$ such that

$$T(1_E) = \lambda(E) = \int_X 1_E \, g \, d\mu$$

for all $E \in \mathfrak{A}$. It follows easily that for each $\phi \in \mathfrak{S}_0$, we have

$$T(\phi) = \int_X \phi \, g \, d\mu.$$

Now let $f \in L^p$, and we suppose without loss of generality that f is nonnegative as well. Then there exists a sequence $\phi_n \in \mathfrak{S}_0$ which increases towards f as a pointwise limit almost everywhere. It follows that $\|f - \phi_n\|_p \to 0$ as $n \to \infty$. Since T is continuous, we have

$$\begin{aligned}
T(f) &= \lim_n T(\phi_n) \\
&= \lim_n \int_X \phi_n \, g \, d\mu \\
&= \int_X f g \, d\mu.
\end{aligned}$$

Thus $T = T_g$ as claimed.

Next, we consider the general case of complex-valued linear functionals. If we restrict T to real L^p-functions, we can write $T = \Re(T) + i\Im(T)$. We apply each of these two real-valued parts separately to real-valued functions f. This produces two results, g_1 and g_2, as in the first part of this proof. Then we let $g = g_1 + ig_2$, and we see that $T = T_g$, a bounded $\mathbb{C}$-linear functional acting on complex L^p-functions.

We need to prove that $g \in L^q$, and that $\|T_g\| = \|g\|_q$. We define a *truncated* function f_n in such a way that $f_n g$ approximates $|g|^q$ from below as follows:

$$f_n(x) = \begin{cases} |g(x)|^{q-1} \operatorname{sgn} g(x) & \text{if } |g(x)|^{q-1} \leqslant n \\ n \operatorname{sgn} g(x) & \text{if } |g(x)|^{q-1} > n \end{cases}$$

for each $n \in \mathbb{N}$.[87] Being bounded on a space of finite measure, f_n lies in L^p for each $n \in \mathbb{N}$. Thus

$$|T(f_n)| = \left| \int_X f_n g \, d\mu \right| \leqslant \|T\| \|f_n\|_p.$$

---

[87]Here the signum function is understood to mean $\operatorname{sgn}(z) = \dfrac{\bar{z}}{|z|}$ if $z \neq 0$, or 0 otherwise.

But

$$f_n g = |f_n||g| \geq |f_n||f_n|^{\frac{1}{q-1}} = |f_n|^p.$$

Thus

$$\int_X |f_n|^p \, d\mu \leq \|T\| \left(\int_X |f_n|^p \, d\mu \right)^{\frac{1}{p}}.$$

Hence

$$\|T\| \geq \left(\int_X |f_n|^p \, d\mu \right)^{1 - \frac{1}{p}}$$

$$= \left(\int_X |f_n|^p \, d\mu \right)^{\frac{1}{q}} \to \|g\|_q.$$

The opposite inequality is contained in Exercise 9.9. This completes the proof of the first case.

2. Suppose again that $\mu(X) < \infty$ but that $p = 1$, so that $q = \infty$.

We obtain $Tf = \int_X fg \, d\mu$ as before, with $g \in L^1$. We need to show that $g \in L^\infty$, which will be contained in L^1 since $\mu(X) < \infty$. Suppose this were false. Then for each $K > 0$ the set

$$A_K = \{x \mid |g(x)| > K\}$$

has strictly positive measure. Define

$$f_K = \frac{1}{\mu(A_K)} 1_{A_K} \operatorname{sgn} g$$

so that $\|f_K\|_1 = 1$, ensuring that $f_K \in L^1(X, \mathfrak{A}, \mu)$. Then we would have

$$|T(f_K)| = \int_X f_K g \, d\mu$$

$$= \int_{A_K} \frac{|g|}{\mu(A_K)} \, d\mu > K$$

for all $K > 0$. This contradicts the boundedness of T. Hence $g \in L^\infty$ as claimed. The reader will complete the proof in Exercise 9.11.

3. Suppose $\mu(X) = \infty$. Since the underlying measure space is σ-finite, we have

$$X = \bigcup_1^\infty A_k,$$

where $A_k \subseteq A_{k+1}$ for all k, and $\mu(A_k) < \infty$.

In this case

$$T : f \big|_{A_k} = \int_X f \big|_{A_k} g_k \, d\mu.$$

This determines $g_k \in L^q(A_k)$ uniquely almost everywhere in such a way that

$$\|g_k\|_q \leqslant \|T\| \tag{9.3}$$

for each $k \in \mathbb{N}$. It is clear that g_k *extends* g_{k-1} for each k, so that $g = \lim_k g_k$ exists pointwise. By Exercise 9.12 we know that

$$T(f) = \int_X f g \, d\mu.$$

We need to show that $g \in L^q$. The functions $|g_k|$ increase toward the limit $|g|$. Thus

$$\int_X |g|^q \, d\mu = \infty \Leftrightarrow \lim_k \int_X |g_k|^q \, d\mu = \infty$$

because of the Monotone Convergence theorem (Theorem 5.4.1). Thus

$$\int_X |g|^q \, d\mu < \infty$$

if $q < \infty$ because of Inequality (9.3).

If $q = \infty$, we have $\|g_k\|_\infty < \|T\|$ for all k. Thus $\|g\|_\infty < \infty$, since the union of countably many null sets is a null set.

∎

EXERCISES

9.11   Complete the proof of case 2 of Theorem 9.3.1 by proving that $\|T\| = \|g\|_\infty$.

9.12   Complete the proof of case 3 of Theorem 9.3.1 by proving that $T(f) = \int_X f g \, d\mu$ for all $f \in L^p(X, \mathfrak{A}, \mu)$.

9.13   Suppose f and g are in $L^p(X, \mathfrak{A}, \mu)$, as in Theorem 9.3.1. If $\dfrac{1}{p} + \dfrac{1}{q} = 1$ and if

$$\int_X fh \, d\mu = \int_X gh \, d\mu$$

for all $h \in L^q(X, \mathfrak{A}, \mu)$, prove that $f = g$ almost everywhere. (See Example 9.1.)

9.4   HILBERT SPACE, ITS DUAL, AND $L^2(X, \mathfrak{A}, \mu)$

Let $(X, \mathfrak{A}, \mu)$ be a σ-finite measure space. According to Theorem 9.3.1, the space $L^2(X, \mathfrak{A}, \mu)$ can be identified with its own dual space, since $\frac{1}{2} + \frac{1}{2} = 1$. The identification is made as follows. If $T \in (L^2)'$, then there exists $g \in L^2$ such that

$$T(f) = \int_X fg \, d\mu$$

for all $f \in L^2$. The equivalence class of g is uniquely determined by T. Moreover,

$$\|f\|_2^2 = \int_X |f|^2 \, d\mu$$
$$= \int_X f\bar{f} \, d\mu$$
$$= \langle f, f \rangle,$$

where the Hermitian scalar product

$$\langle f, g \rangle = \int_X f\bar{g} \, d\mu,$$

for all f and g in L^2.[88] We introduce a definition.

Definition 9.4.1 In any vector space V over the scalar field $\mathbb{C}$ of complex numbers, we call a function

$$\langle \cdot, \cdot \rangle : V \times V \to \mathbb{C}$$

a *Hermitian scalar product*[89] if and only if it has the following three properties:[90]

i. $\langle a\mathbf{x} + \mathbf{y}, \mathbf{z} \rangle = a \langle \mathbf{x}, \mathbf{z} \rangle + \langle \mathbf{y}, \mathbf{z} \rangle$ for all $a \in \mathbb{C}$ and for all $\mathbf{x}$ and $\mathbf{y}$ in V. (*Linearity in the First Variable*)

ii. $\langle \mathbf{x}, \mathbf{y} \rangle = \overline{\langle \mathbf{y}, \mathbf{x} \rangle}$ for all $\mathbf{x}$ and $\mathbf{y}$ in V. (*Conjugate Symmetry*)

iii. $\langle \mathbf{x}, \mathbf{x} \rangle \geqslant 0$ for all $\mathbf{x} \in V$, and $\langle \mathbf{x}, \mathbf{x} \rangle = 0 \Leftrightarrow \mathbf{x} = \mathbf{0} \in V$. (*Positive Definiteness*)

A vector space V, equipped with a Hermitian inner product, is called a *Hermitian inner product space*. If a Hermitian inner product space $\mathcal{H}$ is *complete* with respect to the corresponding norm, as defined in Theorem 9.4.1, then it is called a *Hilbert space*.

Theorem 9.4.1 *In a complex vector space V equipped with a Hermitian scalar product, we define*

$$\|\mathbf{x}\| = \sqrt{\langle \mathbf{x}, \mathbf{x} \rangle}, \tag{9.4}$$

for all vectors $\mathbf{x} \in V$.[91] The function $\| \cdot \|$ as defined in Equation (9.4) is a norm, as in Definition 5.5.1, where scalars c are taken as complex and $|c|$ is interpreted as the modulus of c. Moreover the Cauchy-Schwarz Inequality *is satisfied:*

$$|\langle \mathbf{x}, \mathbf{y} \rangle| \leqslant \|\mathbf{x}\| \|\mathbf{y}\|.$$

---

[88]We express the scalar product here in the proper form for complex-valued functions. If we were dealing only with real-valued functions, then the conjugation sign could be omitted, and the scalar product would be symmetric—not conjugate symmetric.

[89]Another name for this is *Hermitian inner product*.

[90]Part of this section is adapted from [20].

[91]Not all norms in vector spaces can be given by a scalar product. See Exercise 9.18.

Proof: To prove the Cauchy-Schwarz inequality, we fix $\mathbf{x}$ and $\mathbf{y}$, and we proceed as follows. If $\mathbf{x} = \mathbf{0}$, the Cauchy-Schwarz inequality is trivial. So, suppose $\mathbf{x} \neq \mathbf{0}$. For all $c \in \mathbb{C}$, observe that $\langle c\mathbf{x} + \mathbf{y}, c\mathbf{x} + \mathbf{y} \rangle \geq 0$ for all c. By linearity of the scalar product in the first variable and conjugate linearity in the second variable, we see that

$$\|\mathbf{x}\|^2 |c|^2 + 2\Re(c\langle \mathbf{x}, \mathbf{y} \rangle) + \|\mathbf{y}\|^2 \geq 0$$

for all $c \in \mathbb{C}$. (If $z = a + ib \in \mathbb{C}$, we denote the real part of z by $\Re z = a$ and the imaginary part of z by $\Im z = b$.) Let

$$c = -\frac{\langle \mathbf{x}, \mathbf{y} \rangle}{\|\mathbf{x}\|^2}.$$

An easy calculation shows that $|\langle \mathbf{x}, \mathbf{y} \rangle|^2 \leq \|\mathbf{x}\|^2 \|\mathbf{y}\|^2$.

The first two conditions of Definition 5.5.1 are easily verified for $\|\cdot\|$. The third condition, the triangle inequality, is left for Exercise 9.14.   ∎

Corollary 9.4.1 *If V is a Hermitian inner product space, then the mapping T_w, defined by*

$$T_w(v) = \langle v, w \rangle$$

for each $v \in V$, is a bounded linear functional on V.

Proof: This follows immediately from the Cauchy-Schwarz inequality.   ∎

Theorem 9.4.2 *Let $\mathcal{H}$ be any Hilbert space. If a mapping T of a Hilbert Space $\mathcal{H}$ to the scalar field is a bounded linear functional, then there exists a unique element $w \in \mathcal{H}$ such that*

$$T(v) = \langle v, w \rangle$$

for each $v \in \mathcal{H}$. Thus the dual space of $\mathcal{H}$ is identified with $\mathcal{H}$ itself.

Proof: If $T \equiv 0$, then we can take $w = 0$, and that is the only choice that suffices. Suppose $T \neq 0 \in \mathcal{H}'$. It is easy to see that the $\ker(T)$, the *kernel* of the linear transformation T, is a closed, proper subspace of $\mathcal{H}$. Moreover, if $x \notin \ker(T)$, then any other such vector x' must be congruent to x modulo $\ker(T)$. Thus the codimension of $\ker(T)$ is 1, and $\mathcal{H} = \ker(T) \oplus \mathbb{C}x$.

Suppose for the moment that we can choose x to be a *unit* vector orthogonal to $\ker(T)$ in the sense of the Hermitian scalar product of x with any vector from $\ker(T)$ being zero. Then we can check readily that

$$T(v) \equiv \left\langle v, \overline{T(x)}x \right\rangle,$$

so that we can take $w = \overline{T(x)}x$.[92] The reader should verify that this mapping is well defined, since a unit vector orthogonal to the $\ker(T)$ is defined only up to a complex

---

[92] The choice of the second variable of the scalar product as the proper location for the vector that represents the action of T on v assures that the action on v is linear, in the environment of a complex Hilbert space.

scalar factor of modulus one. It remains only to show that if V is a closed, proper, nontrivial subspace of $\mathcal{H}$, then there exists a nonzero vector $x \in V^{\perp}$, the subspace of all vectors orthogonal to each vector in V. The latter space is known also as the *orthogonal complement* of V. The reader will prove this in Exercise 9.19. ∎

Remark 9.4.1 The map that carries $y \to T_y$ is an isomorphism of $\mathcal{H}$ onto its dual $\mathcal{H}'$ if $\mathcal{H}$ is a Hilbert space over the field $\mathbb{R}$ but not if the field is $\mathbb{C}$. It is easy to check that the mapping $y \to T_y$ is linear over the real field. But over the complex field $T_{cy} = \bar{c}T_y$, which causes the map $y \to T_y$ to fail to be linear. If $\mathcal{H}$ happens to be $L^2(X, \mathfrak{A}, \mu)$, then over either field, $y \to T_y$ preserves the Hilbert space norm because of Theorem 9.3.1. The latter theorem does provide a *Banach space* isomorphism between L^2 and its dual. But in the present discussion we are working in the context of an *abstract*, complex Hilbert space, and for this reason T_y is being defined by means of the Hermitian scalar product.

We can make $\mathcal{H}'$ into a Hilbert space in its own right, by introducing the following scalar product on $\mathcal{H}'$:

$$\langle T_y, T_z \rangle_{\mathcal{H}'} = \langle z, y \rangle.$$

The reader should check easily that $\langle \cdot, \cdot \rangle_{\mathcal{H}'}$ is a Hermitian scalar product.[93]

Definition 9.4.2 An *orthonormal* subset $E = \{e_\alpha \mid \alpha \in A\} \subset \mathcal{H}$, a Hermitian inner product space, is a set with the property that

$$\langle e_\alpha, e_\beta \rangle = \begin{cases} 1 & \text{if } \alpha = \beta, \\ 0 & \text{if } \alpha \neq \beta. \end{cases}$$

The set A is called an *index* set. If $f \in \mathcal{H}$, a Hilbert space with an orthonormal set indexed by A, then we define an abstract *Fourier transform* $\hat{f} : A \to \mathbb{C}$ by

$$\hat{f}(\alpha) = \langle f, e_\alpha \rangle.$$

The reader should note that the Fourier transform, as a function on the index set A, is *dependent* upon the choice of an indexed orthonormal set E.

Observe that if $\hat{f}(\alpha) \neq 0$ for uncountably many values of $\alpha \in A$, then there could be *no finite upper bound* on the set of all possible finite sums of squares of the form $|f(\alpha)|^2$. The next theorem will establish that for each $f \in \mathcal{H}$, we can have $\hat{f}(\alpha) \neq 0$ for at most countably many values of α.

Theorem 9.4.3 (Bessel's Inequality) *Let* $f \in \mathcal{H}$, *a Hilbert space with an orthonormal set E, indexed by A. Then for each* finite *set $F \subseteq A$, we have Bessel's inequality:*

$$\sum_{\alpha \in F} \left| \hat{f}(\alpha) \right|^2 \leq \|f\|^2, \tag{9.5}$$

---

[93]One can interpret the scalar product $\langle \cdot, \cdot \rangle_{\mathcal{H}'}$ as providing an isomporphism of the Hilbert space $\mathcal{H}$ onto its own *second* dual space, $\mathcal{H}'' = (\mathcal{H}')'$. A Hilbert space is an example of a Banach space B for which B'' is isomorphic to B. Such Banach spaces are called *reflexive*. It is not hard to see that Theorem 9.3.1 establishes the reflexivity of the Banach spaces $L^p(X, \mathfrak{A}, \mu)$, provided that $1 < p < \infty$.

with the right side being necessarily finite. Moreover, we have

$$\left\| f - \sum_{\alpha \in F} \widehat{f}(\alpha) e_\alpha \right\|^2 = \|f\|^2 - \sum_{\alpha \in F} \left| \widehat{f}(\alpha) \right|^2.$$

Proof: We calculate *carefully* the following nonnegative Hermitian scalar product:

$$\left\langle f - \sum_{\alpha \in F} \widehat{f}(\alpha) e_\alpha, f - \sum_{\alpha \in F} \widehat{f}(\alpha) e_\alpha \right\rangle = \|f\|^2 - \sum_{\alpha \in F} \overline{\widehat{f}(\alpha)} \widehat{f}(\alpha)$$

$$= \|f\|^2 - \sum_{\alpha \in F} \left| \widehat{f}(\alpha) \right|^2 \geqslant 0.$$

Thus the finite partial sums of the terms $\left| \widehat{f}(\alpha) \right|^2$ are all bounded above by $\|f\|^2$, and this yields Bessel's inequality.   ∎

Definition 9.4.3   A Hilbert space is called *separable* provided that it has a *countable* orthonormal subset $E = \{e_n \mid n \in \mathbb{N}\}$, with respect to which the *Plancherel identity* is satisfied:

$$\|f\|^2 = \sum_{n \in \mathbb{N}} \left| \widehat{f}(n) \right|^2, \tag{9.6}$$

for each $f \in \mathcal{H}$. The subset E is then called a *countable orthonormal basis* for $\mathcal{H}$.

Theorem 9.4.4   *Let $\mathcal{H}$ be a separable Hilbert space with an orthonormal basis $E = \{e_n\}$, indexed by $\mathbb{N}$. Then we have the following conclusion for each $f \in \mathcal{H}$. We have*

$$f = \sum_{n \in \mathbb{N}} c_n e_n,$$

in the sense of convergence with respect to the Hilbert space norm, if and only if each complex coefficient $c_n = \widehat{f}(n)$.

Proof: In one direction the corollary follows immediately from Bessel's inequality. For the uniqueness of the coefficients c_n, expand $\langle f, e_n \rangle$ as an infinite series, using the continuity of the bounded linear functional on $\mathcal{H}$ that is determined by e_n with respect to the Hilbert space norm.[94]   ∎

The reader will see in Exercise 9.15 that a Hilbert space is separable if and only if it has a countable, dense subset. The reader will show also that $L^2(\mathbb{R})$ is a separable Hilbert space.

It is a consequence of Bessel's inequality that for each element f of a separable Hilbert space $\mathcal{H}$, $\widehat{f}$ lies in l_2, the separable Hilbert space of square summable sequences, which the reader will study in Exercise 9.16. Thus the Fourier transform as

---

[94]Consider the Cauchy-Schwarz inequality.

an *operator* can be viewed as a mapping

$$\hat{} : \mathcal{H} \to l_2.$$

The reader will show in Exercise 9.17 that the Fourier transform is a *Hilbert space isomorphism* from $\mathcal{H}$ to l_2. Thus *all separable Hilbert spaces are isomorphic to one another.*

EXERCISES

9.14   If $\|\mathbf{x}\|$ is defined by means of a Hermitian inner product, prove the *triangle inequality*:

$$\|\mathbf{x} + \mathbf{y}\| \leqslant \|\mathbf{x}\| + \|\mathbf{y}\|.$$

(Hint: Use the Cauchy-Schwarz inequality.)

9.15   Let $\mathcal{H}$ be a Hilbert space.

  a) Prove that $\mathcal{H}$ is separable if and only if it possesses a countable, dense subset. That is, $\mathcal{H}$ is separable if and only if there is a countable subset S, the *closure* $\bar{S}$ of which is the whole space $\mathcal{H}$. (For the proof of sufficiency, use a countable iteration of the Gram-Schmidt orthonormalization [95] process from linear algebra.)

  b) Prove that $L^2(\mathbb{R})$ is a separable Hilbert space. (Hint: Let the set $S_{\mathbb{Q}}$ be the set of all step functions σ with rational values, and with the property that $\sigma^{-1}(q)$ is an interval with rational endpoints for each $q \in \mathbb{Q}$. Show that $S_{\mathbb{Q}}$ is dense in $L^2(\mathbb{R})$.)

9.16   Show that the square-summable sequence space

$$l_2 = \left\{ z = (z_1, z_2, \dots, z_n, \dots) \,\middle|\, \sum_{n \in \mathbb{N}} |z_n|^2 < \infty \right\}$$

is a separable Hilbert space. (Hint: Show that $l_2 = L^2(X, \mathfrak{A}, \nu)$ for a suitable choice of σ-finite measure space $(X, \mathfrak{A}, \nu)$.)

9.17   We show in this exercise that all separable Hilbert spaces are isomorphic to l_2.

  a) Prove the following generalization of the Plancherel identity, called *Parseval's identity*: for each f and g in a separable Hilbert space $\mathcal{H}$, we have [96]

$$\langle f, g \rangle = \langle \hat{f}, \hat{g} \rangle.$$

  Parseval's identity shows that the Fourier transform preserves the Hermitian scalar product. (Hint: Apply the Plancherel identity to $\langle f + g, f + g \rangle$ to

---

[95] See, for example, [12].
[96] In a separable Hilbert space, we assume that a countable orthonormal basis has been chosen, and that the abstract Fourier transform has been defined with respect to that basis.

prove that

$$\Re\langle f, g\rangle = \Re\langle \hat{f}, \hat{g}\rangle.$$

Then do something similar with if to obtain the desired conclusion.)

b) Prove that the Fourier transform $\hat{}$ is an isomorphism of a separable Hilbert space $\mathcal{H}$ onto l_2. That is, show that the Fourier transform is a linear bijection that preserves the scalar product and the norm.

9.18 Let V be a real or complex vector space.

a) Suppose V is equipped with an inner product $\langle \cdot, \cdot \rangle$, and suppose we define a corresponding norm by $\|\mathbf{x}\|^2 = \langle \mathbf{x}, \mathbf{x} \rangle$. Prove the *Parallelogram Law:*

$$\|\mathbf{x} + \mathbf{y}\|^2 + \|\mathbf{x} - \mathbf{y}\|^2 = 2\|\mathbf{x}\|^2 + 2\|\mathbf{y}\|^2.$$

b) Prove that the taxicab norm,[97] $\|\mathbf{x}\|_t = |x_1| + |x_2|$, does not correspond, as in 9.18.a, to any inner product on $\mathbb{R}^2$.

c) Under the hypotheses of 9.18.a, prove the identity

$$\Re\langle \mathbf{x}, \mathbf{y} \rangle = \frac{1}{4} \left(\|\mathbf{x} + \mathbf{y}\|^2 - \|\mathbf{x} - \mathbf{y}\|^2 \right),$$

with the understanding that in a real inner product space the real part of the scalar product is the inner product. In a Hermitian inner product space, show that one can express $\Im\langle \mathbf{x}, \mathbf{y} \rangle = -\Re\langle i\mathbf{x}, \mathbf{y} \rangle$.

d) Suppose only that V is a real or complex vector space with a norm. Define what is *hoped* to be a scalar product on V by the formulas in 9.18.c. Prove that this defines a legitimate scalar product[98] on V, provided that the norm satisfies the Parallelogram Law of 9.18.a.

9.19 Prove that if V is a closed, proper, nontrivial subspace of an arbitrary Hilbert space $\mathcal{H}$, then there exists a nontrivial proper closed subspace $V^\perp$, each vector of which is orthogonal to each vector of V, and $\mathcal{H} = V \oplus V^\perp$. This can be done by the following sequence of steps.

a) Let $x \notin V$ and let $d = \inf\{\|x - y\| \mid y \in V\}$. Show that $d > 0$.

b) Pick a sequence $y_n \in V$ such that $\|x - y_n\| \to 0$ as $n \to \infty$. Apply the Parallelogram Law (from Exercise 9.18.a) to the sequences $x - y_n$ and $x - y_m$, to prove that y_n is a Cauchy sequence.

c) Prove that $y = \lim_{n \to \infty} y_n$ has the property that $x - y$ is orthogonal to V and is nonzero.

d) Denote the orthogonal complement of V by

$$V^\perp = \{w \in \mathcal{H} \mid \langle v, w \rangle = 0 \ \forall v \in V\}.$$

Prove that $V^\perp$ is a closed subspace of $\mathcal{H}$ and that

$$\mathcal{H} = V \oplus V^\perp.$$

---

[97]The taxicab norm is defined in many advanced calculus books, such as [20].
[98]For this and generalizations, see [13].

9.5 RIESZ-MARKOV-SAKS-KAKUTANI THEOREM

Early in the twentieth century, Frigyes Riesz discovered a full classification (or *representation*) of the dual space for the vector space $C[a, b]$, consisting of the continuous functions on $[a, b]$ and equipped with the L^{∞}-norm, or sup-norm. This space, with the given norm, is a complete normed linear space: It is a Banach space in modern terminology. Riesz proved that each bounded linear functional on that space can be described as

$$T(f) = \int_a^b f \, d\mu,$$

using a suitable finite Borel measure μ.[99] Later, Andrei A. Markov extended this theorem to the compactly supported functions on the infinite real line. A version was proven by Stanislaw Saks[100] for $C(X)$, with the hypothesis that X is a compact metric space. And Shizuo Kakutani[101] generalized the theorem to cover the vector space of all continuous functions on any compact Hausdorff space. *We will prove the Kakutani version of the theorem.* The proof we give is adapted slightly from the one presented by Kakutani.[102]

In order to be able to integrate continuous functions on a compact topological space, it is natural to use the field generated by either the open or the closed subsets for elementary sets. It will make no difference which is chosen since we are dealing with a field of sets. Either way, a continuous function will be measurable.

Definition 9.5.1 A measure μ on the Borel field $\mathfrak{B}$ generated by the open subsets of a compact space is called *regular*, provided that for each $E \in \mathfrak{B}$ and for each $\epsilon > 0$, we have a closed set $F \subseteq E$ and an open set $G \supseteq E$ such that

$$\mu(G) - \mu(F) < \epsilon.$$

If μ is a *signed Borel measure*, then we call μ regular provided that the variation $|\mu|$ (from Definition 8.1.2) be regular, under the definition for positive Borel measures. This is equivalent to both μ^+ and μ^- being regular.

Theorem 9.5.1 (Riesz-Markov-Saks-Kakutani) *Let X be any compact Hausdorff space, and let $C(X)$ be the vector space of all continuous real-valued functions on X, equipped with the L^{∞}-norm. Then the dual space $C(X)'$ is isomorphic as a Banach space to the space $\mathcal{M}(X)$ of all finite,* regular *signed measures defined on the Borel field generated by the open (or, equivalently, the closed) sets. The space*

---

[99]Actually, Riesz expressed this representation as a Stieltjes integral with respect to a function of bounded variation. See, for example, [21] or [20].

[100]See [23].

[101]Kakutani produced this theorem as part of a paper [14] that provides a classification of all objects known as Banach lattices.

[102]The author is calling the theorem after all four mathematicians of whose contributions he is aware. However, it is common to see this theorem named after the first two of these authors.

$\mathcal{M}(X)$ *will be equipped with the total variation norm, denoted by* $\|\mu\|$. [103] *The correspondence is given by*

$$T_\mu(f) = \int_X f \, d\mu.$$

Proof:

i. We will begin by proving that the map from $\mu \to T_\mu$ is *both* isometric and injective. Let $\mu \in \mathcal{M}(X)$ with norm $\|\mu\|$, as in Exercise 8.3. Define

$$T_\mu(f) = \int_X f \, d\mu.$$

We see that $|T_\mu(f)| \leq \|f\|_x \|\mu\|$, so that

$$\|T_\mu\| \leq \|\mu\|.$$

We would like to show equality, however, for these two norms. For this purpose, write the Hahn decomposition $X = P \cup N$, as in Theorem 8.1.2. Let $\epsilon > 0$. There exist compact sets K_P and K_N and open sets O_P and O_N such that

$$K_P \subseteq P \subseteq O_P \text{ and } K_N \subseteq N \subseteq O_N$$

and such that

$$|\mu|(O_P \backslash K_P) < \epsilon \text{ and } |\mu|(O_N \backslash K_N) < \epsilon.$$

It follows from Urysohn's lemma that there exists a continuous, real-valued function f with $\|f\|_\infty = 1$ such that

$$f : K_P \backslash O_N \to 1 \text{ and } f : K_N \backslash O_P \to -1.$$

Observe that

$$|\mu|(K_P \backslash O_N) \geq \mu(P) - 2\epsilon$$

and

$$|\mu|(K_N \backslash O_P) \geq |\mu|(N) - 2\epsilon.$$

Thus

$$\left| \int_X f \, d\mu \right| \geq \|\mu\| - 8\epsilon$$

for each $\epsilon > 0$. Therefore $\|T\| \geq \|\mu\|$, making both norms equal.

Suppose now that T could be expressed by two regular Borel measures, μ and ν. Then $T - T = 0$ would correspond to $\mu - \nu$, and this together with the first part of the proof implies that $\|\mu - \nu\| = 0$. Hence $\mu = \nu$. It follows that the mapping from $\mathcal{M} \to \mathcal{C}(X)'$ is an injection.

---

[103] See Definition 8.1.2 for $\|\mu\|$.

Remark 9.5.1 The main part of the proof is to show that the injection, $\mu \to T_\mu$, is also a surjection. The reason this is difficult to prove is that we are given a bounded linear map $T : C(X) \to \mathbb{R}$, and we would like to determine a measure that could describe T in the manner given above. The first guess could be that if $E \in \mathfrak{B}$, we should let $\mu(E) = T(1_B)$. This would not make sense, however, because the function 1_B is usually not continuous, for which reason $T(1_B)$ is undefined.

ii. We will show how we can construct a suitable measure, given a bounded linear functional T, with the temporary simplifying assumption that T *is a positive operator*, as in the following definition.

Definition 9.5.2 A bounded linear functional $T \in C(X)'$ is called *positive* if and only if $Tf \geqslant 0$ for each function f that is everywhere nonnegative.

The reader will need to prove, in Exercise 9.20, that each positive linear functional is also monotone, as explained in that exercise. The restriction to positive operators will be lifted in the final part of the proof.

Since T has a positive, finite norm, *we can assume that* $\|T\| = 1$ simply by adjusting T by a scalar factor.[104] *We make this assumption for simplicity in what follows.*

We can use T to define a nonnegative function μ on the family $\mathfrak{O}$ of all open sets by

$$\mu(O) = \sup \{Tf \mid 0 \leqslant f \leqslant 1_O, \ f \in C(X)\}. \tag{9.7}$$

Here 1_O is the indicator function of the set O.

Lemma 9.5.1 *The set function μ has the following properties on the set $\mathfrak{O}$ of all open sets:*

(a) (Null Set Additivity) $\mu(\varnothing) = 0$.

(b) (Monotonicity) $O_1 \subset O_2$ *implies that* $\mu(O_1) \leqslant \mu(O_2)$.

(c) (Subadditivity) $\mu(O_1 \cup O_2) \leqslant \mu(O_1) + \mu(O_2)$.

Proof: The first two parts we leave as an informal exercise for the reader. For the third part we reason as follows.

Let $\epsilon > 0$. Then there exists $f \in C(X)$ with $0 \leqslant f \leqslant 1_{O_1 \cup O_2}$ such that

$$Tf > \mu(O_1 \cup O_2) - \epsilon.$$

Let $C = f^{-1}[\epsilon, 1]$, so that C is a compact subset of $O_1 \cup O_2$. We claim that we can write C as the (not necessarily disjoint) union of closed (and thus compact) sets, C_1 and C_2, such that $C_1 \subset O_1$ and $C_2 \subset O_2$.

---

[104]We assume that $T \neq 0$, the zero operator, since Theorem 9.5.1 would be trivial in that case.

In fact, the boundary set $\partial(O_1 \cap C)\backslash O_1$ and the boundary set $\partial(O_2 \cap C)\backslash O_2$ are mutually disjoint subsets of $C \cap O_2$ and $C \cap O_1$, respectively. Since the space X is both compact and Hausdorff, there exist mutually disjoint open sets $U_2 \subseteq O_2$ and $U_1 \subseteq O_1$ such that

$$\partial(O_1 \cap C)\backslash O_1 \subseteq U_2,$$

and

$$\partial(O_2 \cap C)\backslash O_2 \subseteq U_1.$$

We can select $C_1 = C \cap O_1 \backslash U_2$ and $C_2 = C \cap O_2 \backslash U_1$.

Thus there exist continuous functions f_1 and f_2 such that $f_1 \equiv 1$ on C_1 and $0 \leqslant f_1 \leqslant 1_{O_1}$, and $f_2 \equiv 1$ on C_2 and $0 \leqslant f_2 \leqslant 1_{O_2}$. It follows from the definition of the set C that

$$f \leqslant f_1 + f_2 + \epsilon.$$

Also, $T(f_1) \leqslant \mu(O_1)$ and $T(f_2) \leqslant \mu(O_2)$. It follows that

$$
\begin{aligned}
\mu(O_1 \cup O_2) - \epsilon &\leqslant T(f) \qquad\qquad\qquad\qquad (9.8) \\
&\leqslant T(f_1) + T(f_2) + T(\epsilon) \\
&\leqslant \mu(O_1) + \mu(O_2) + \epsilon,
\end{aligned}
$$

since we are assuming that $\|T\| = 1$. Because the inequalities (9.8) hold for all $\epsilon > 0$, the lemma has been proven. ∎

Our next task is to extend the domain of μ to $\mathfrak{P}(X)$, which we do as follows. We let

$$\mu^*(E) = \inf\{\mu(O) \mid O \supseteq E, \ O \in \mathfrak{O}\}, \qquad\qquad (9.9)$$

for each subset E of X. Observe that for an open set O we have $\mu(O) = \mu^*(O)$.

Lemma 9.5.2 *The function μ^* is a Carathéodory outer measure on the σ-field $\mathfrak{P}(X)$, the power set of X.*

Proof: We need to show that $\mu^*(\varnothing) = 0$, and that μ^* is monotone and countably subadditive. The first requirement is a simple consequence of the fact that $\varnothing$ is open, and $1_\varnothing \leqslant 0$ on X. Monotonicity follows immediately from the definition of μ^* as an infimum. Finally, if $E_i \in \mathfrak{P}(X)$ for each $i \in \mathbb{N}$, let $\epsilon > 0$, and take an open set $O_i \supseteq E_i$ such that

$$\mu(O_i) < \mu^*(E_i) + \frac{\epsilon}{2^i}.$$

It follows from Lemma 9.5.1 that for each $n \in \mathbb{N}$ we have

$$\mu^*\left(\bigcup_{i=1}^{n} E_i\right) \leqslant \mu\left(\bigcup_{i=1}^{n} O_i\right)$$

$$\leqslant \sum_{i=1}^{n} \mu(O_i)$$

$$< \sum_{i=1}^{n} \mu^*(E_i) + \epsilon$$

$$\leqslant \sum_{i=1}^{\infty} \mu^*(E_i) + \epsilon.$$

Since this is true for all $\epsilon > 0$,

$$\mu^*\left(\bigcup_{i\in\mathbb{N}} E_i\right) \leqslant \sum_{i\in\mathbb{N}} \mu^*(E_i),$$

and this proves the lemma. ∎

The next lemma will imply that each Borel set is μ^*-measurable.

Lemma 9.5.3 *Each open set O is μ^*-measurable.*

Proof: Since we know that μ^* is subadditive, it will suffice to prove for each $E \subset X$ that

$$\mu^*(E) \geqslant \mu^*(E \cap O) + \mu^*(E \cap O^c).$$

As in the previous lemma, Equation (9.9) tells us that it will suffice to prove the inequality for every *open* set E. To prove this, we proceed as follows. Let $\epsilon > 0$. Then there exists $f \in C(X)$ with $0 \leqslant f \leqslant 1_{E \cap O}$ such that

$$Tf > \mu(E \cap O) - \epsilon.$$

Let $C = f^{-1}[\epsilon, 1]$, which is a closed subset of $E \cap O$. Similarly, there exists $g \in C(X)$ with $0 \leqslant g \leqslant 1_{E \backslash C}$ and $Tg > \mu(E \backslash C) - \epsilon$. Thus

$$0 \leqslant f + g \leqslant 1_E + \epsilon,$$

and this implies that $Tf + Tg \leqslant \mu(E) + \epsilon$. Thus

$$\mu^*(E \cap O^c) + \mu^*(E \cap O) - 2\epsilon \leqslant \mu(E \backslash C) + \mu(E \cap O) - 2\epsilon$$
$$< Tf + Tg$$
$$< \mu(E) + \epsilon,$$

which proves the lemma since $\epsilon > 0$ is arbitrary. ∎

That μ^* is a regular Carathéodory outer measure follows from Exercise 2.20. It follows that μ^* defines a regular measure on the σ-algebra of all μ^*-measurable subsets of X.

Lemma 9.5.4 *For each $f \in \mathcal{C}(X)$, we have $\int_X f \, d\mu = Tf$.*

Proof: We may assume, without loss of generality, that f is not the constant zero function. Let $\epsilon > 0$, and pick a finite sequence of numbers α_i such that

$$\alpha_0 < \alpha_1 < \cdots < \alpha_n$$

and such that $\alpha_0 < -\|f\|_\infty < \|f\|_\infty < \alpha_n$ and $\alpha_i - \alpha_{i-1} < \epsilon$ for each i. Define the open set

$$E_i = f^{-1}(\alpha_i, \infty),$$

for $i = 1, 2, \ldots, n$. Define the functions

$$f_i(t) = \begin{cases} 0 & \text{if } f(t) < \alpha_{i-1}, \\ \frac{f(t) - \alpha_{i-1}}{\alpha_i - \alpha_{i-1}} & \text{if } \alpha_{i-1} \leqslant f(t) \leqslant \alpha_i, \\ 1 & \text{if } f(t) > \alpha_i \end{cases}$$

for $i = 1, \ldots, n$. Thus $f_i(t) \leqslant 1$ if $t \in E_{i-1}$, $f_i(t) = 0$ if $t \in E_{i-1}^c$, and

$$f(t) = \alpha_0 + \sum_{i=1}^n (\alpha_i - \alpha_{i-1}) f_i(t).$$

Hence $T(f_i) \leqslant \mu(E_{i-1})$, by Equation (9.7). Also, $\mu(E_0) = 1$, and $\mu(E_n) = 0$. Thus

$$\begin{aligned}
Tf &= \alpha_0 + \sum_{i=1}^n (\alpha_i - \alpha_{i-1}) T(f_i) && (9.10) \\
&\leqslant \alpha_0 + \sum_{i=1}^n (\alpha_i - \alpha_{i-1}) \mu(E_{i-1}) \\
&= \sum_{i=1}^n \alpha_i \big(\mu(E_{i-1}) - \mu(E_i)\big) \\
&\leqslant \sum_{i=1}^n \alpha_{i-1} \big(\mu(E_{i-1}) - \mu(E_i)\big) + \epsilon \\
&\leqslant \int_X f(t) \, d\mu + \epsilon,
\end{aligned}$$

where we use again the assumption that $\|T\| \leqslant 1$, making $\mu(X) \leqslant 1$ as well. Since the inequalities (9.10) are true for all $\epsilon > 0$ we have $Tf \leqslant \int_X f \, d\mu$. But the opposite inequality follows if we replace f by $-f$ and use the linearity of T. This proves the lemma. ∎

iii. There is only one final step in the proof of Theorem 9.5.1. We need to prove that we can separate a general bounded linear functional $T \in C(X)'$ into the difference of two *positive parts*. This is accomplished in the next lemma, which is sometimes also called a Riesz representation theorem, as in [24], where it is proven in broader generality.

Lemma 9.5.5 *If $T \in C(X)'$, then there exist two* positive *bounded linear functionals, T^+ and T^-, such that*

$$T = T^+ - T^-, \quad and$$
$$\|T\| = T^+(1) + T^-(1).$$

Proof: Note that T is not necessarily positive. Suppose $f \geqslant 0$, and define

$$T^+ f = \sup_{0 \leqslant \phi \leqslant f} T(\phi).$$

It follows that

$$T^+ f \geqslant T f,$$

and $T^+(cf) = cT^+ f$ for all $c > 0$. We will need to prove that T^+ is additive as part of the work of proving that it has a linear extension. Now let f and g both be nonnegative with $0 \leqslant \phi \leqslant f$ and $0 \leqslant \psi \leqslant g$. Thus

$$0 \leqslant \phi + \psi \leqslant f + g,$$

so that $T^+(f + g) \geqslant T\phi + T\psi$. Hence

$$T^+(f + g) \geqslant T^+ f + T^+ g.$$

On the other hand, if $0 \leqslant \psi \leqslant f + g$, then

$$0 \leqslant \psi \wedge f \leqslant f, \quad and$$
$$0 \leqslant \psi - \psi \wedge f \leqslant g.$$

Thus

$$T\psi = T(\psi \wedge f) + T(\psi - \psi \wedge f)$$
$$\leqslant T^+ f + T^+ g.$$

It follows that T^+ is additive, since

$$T^+(f + g) \leqslant T^+ f + T^+ g.$$

Now let $f \in C(X)$ be *arbitrary* (*not* necessarily positive) and pick any two constants M and N such that $f + M \geqslant 0$ and $f + N \geqslant 0$. Thus

$$T^+(f + M + N) = T^+(f + M) + T^+ N$$
$$= T^+(f + N) + T^+ M,$$

which implies that

$$T^+(f + M) - T^+M = T^+(f + N) - T^+N.$$

Hence it makes sense to *define*

$$T^+f = T^+(f + M) - T^+M,$$

for any M such that $f + M \geqslant 0$. The reader should check that T^+ is additive by doing Exercise 9.21.

Observe also that

$$T^+(-f) + T^+f = T^+0 = 0,$$

so that $T^+(-f) = -T^+f$. It follows that T^+ is linear. Hence we can define

$$T^- = T - T^+,$$

so that T^- must be linear, $T^- \geqslant 0$, and $T = T^+ - T^-$.

All that remains is to prove that $\|T\| = T^+(1) + T^-(1)$. Observe that if T were positive, then $f \geqslant g$ would imply that $Tf \geqslant Tg$. But

$$\|T\| = \sup_{f \neq 0} \frac{|Tf|}{\|f\|}$$

and $\left| \dfrac{f}{\|f\|} \right| \leqslant 1$. Thus if $T \geqslant 0$, it follows that $\|T\| = T(1)$. Since for general T we have $T = T^+ - T^-$, it follows that

$$\begin{aligned} \|T\| &\leqslant \|T^+\| + \|T^-\| \\ &= T^+(1) + T^-(1). \end{aligned}$$

On the other hand, if $0 \leqslant \phi \leqslant 1$, then $|2\phi - 1| \leqslant 1$ and

$$\begin{aligned} \|T\| &= \sup_{0 \leqslant \phi \leqslant 1} |T(\phi)| \\ &\geqslant T(2\phi - 1) \\ &= 2T(\phi) - T(1). \end{aligned}$$

Thus

$$\begin{aligned} \|T\| &\geqslant 2T^+(1) - T(1) \\ &= T^+(1) + T^-(1). \end{aligned}$$

It follows that $\|T\| = T^+(1) + T^-(1)$. This proves the lemma. ∎

The proof of Theorem 9.5.1 is complete. ∎

EXERCISES

9.20   Prove that if T is a positive linear functional on $\mathcal{C}(X)$, then T is monotone. That is $f \leqslant g$ implies that $Tf \leqslant Tg$.

9.21   Prove that $T^+(f + g) = T^+f + T^+g$ for arbitrary continuous functions f and g on X, and that $T^+(cf) = cT^+(f)$ for all $c \geqslant 0$.

9.22   Suppose μ is a bounded regular (signed) Borel measure on $[0, 1] \subset \mathbb{R}$, and suppose $\int_0^1 x^n \, d\mu = 0$ for each nonnegative integer n. Prove that $\mu = 0$.

9.23   For each bounded (signed) Borel measure μ defined on the Lebesgue measurable sets of $[0, 1]$, define the *Fourier-Stieltjes transform* to be the function $\hat{\mu} : \mathbb{Z} \to \mathbb{C}$, given by

$$\hat{\mu}(n) = \int_0^1 e^{-2\pi i n x} \, d\mu$$

for each $n \in \mathbb{Z}$. If μ and ν are two bounded Borel measures on $[0, 1]$ for which $\hat{\mu} \equiv \hat{\nu}$, prove that $\mu = \nu$. In words, this exercise states that the Fourier-Stieltjes transform of a bounded Borel measure determines that measure uniquely.

CHAPTER 10

TRANSLATION INVARIANCE IN REAL ANALYSIS

Throughout this book, we have emphasized the example of $\mathbb{R}^n$, with $n \geq 1$. An important attribute of $\mathbb{R}^n$, which distinguishes it from a general measure space, is the fact that $\mathbb{R}^n$ is a group under vector addition. In the case of $\mathbb{R}^n$, Lebesgue measure is translation-invariant, as is the Lebesgue integral.

In this chapter we carry the theme of translation invariance further, classifying the closed, translation-invariant subspaces of $L^2(\mathbb{R})$ and $L^2(\mathbb{T})$, where $\mathbb{T}$ is the unit circle, meaning the quotient group $\mathbb{R}/\mathbb{Z}$. The closed, translation-invariant subspaces will be differentiation-invariant as well, meaning that differentiable functions in these subspaces will have their derivatives in the same subspaces. [105]

We will begin with the decomposition of L^2 of the circle into a direct sum of one-dimensional translation-invariant subspaces $\mathbb{C}e^{2\pi i n x}$ of L^2. Thus each invariant summand in the direct sum will be the space of all complex scalar multiples of the periodic function $e^{2\pi i n x}$. The reader should note that each translate of the latter function is a complex scalar multiple of itself. This will be followed by a treatment of

---

[105]The theorems to be treated in this chapter are special cases of theorems that apply more generally to functions on suitable topological groups or on Lie groups for the consideration of differentiation.

Measure and Integration: A Concise Introduction to Real Analysis. By Leonard F. Richardson
Copyright © 2009 John Wiley & Sons, Inc.

$L^2(\mathbb{R})$, expanded as a *direct integral* of one-dimensional translation-invariant spaces, with n replaced by a real parameter.[106]

In order to develop the decomposition of L^2 of a circle as indicated, it will be helpful to treat the smooth functions first. For $L^2(\mathbb{R})$ we will consider *Schwartz functions* in a similar role.

The present chapter may be interpreted as an introduction to Fourier analysis, or to harmonic analysis, on the real line, but from the perspective of its modern group-theoretic sense.

10.1   AN ORTHONORMAL BASIS FOR $L^2(\mathbb{T})$

We regard the circle as being the quotient group of the additive group of real numbers, modulo the integers, or any nonzero multiple thereof. We will prove most of our theorems in the context of

$$\mathbb{T} = \mathbb{R}/\mathbb{Z},$$

although we will need to generalize this slightly to $\mathbb{R}/(k\mathbb{Z})$,[107] for any nonzero constant $k \in \mathbb{R}$.[108] Sometimes we will need to stress that $\mathbb{T} = \mathbb{R}/\mathbb{Z}$ is being considered as a group under the operation of *addition*, so we will denote the *additive circle group* by

$$(\mathbb{T}, +).$$

In the case of $\mathbb{R}/\mathbb{Z}$, the circle may be identified with the unit interval, with the numbers 0 and 1 identified. Geometrically, this can be modeled by bending the unit interval into a circle and gluing 0 and 1 together. A function

$$f : \mathbb{T} \to \mathbb{C}$$

is understood to mean a periodic function on the real line, with period 1. Thus

$$f(x) \equiv f(x+1),$$

meaning that the equality is valid for each real number x.

Remark 10.1.1 Unless we make a statement to the contrary, $L^2(\mathbb{T})$ is understood to be with respect to Lebesgue measure on the interval $[0, 1)$, with each point of the circle being identified with its unique coset representative from $[0, 1)$, which is regarded also as being a *cross section* for the coset space $\mathbb{R}/\mathbb{Z}$. We leave it as an exercise for the reader[109] to prove that this measure is translation-invariant with respect to the action by addition of the group $\mathbb{R}$ on T. Observe that if $f \in L^2(\mathbb{T})$, then it is true also that $f \in L^1(\mathbb{T})$. This is necessary for the following definition, in

---

[106]The reader should note the absence in the latter case of the prefix *sub*!

[107]See Exercise 10.6.

[108]The symbol $\mathbb{T}$ is appropriate for a circle, because we think of the circle as a one-dimensional *torus*. A product of n circle groups is an n-torus, which we denote by $\mathbb{T}^n$.

[109]See Exercise 10.2.

which we introduce two key concepts: The Fourier transform and the Fourier series of $f \in L^2(\mathbb{T})$.

Definition 10.1.1 In the following statements, $L^2(\mathbb{T})$ and $L^1(\mathbb{T})$ refer to Lebesgue measure as described in Remark 10.1.1.

   i. For each $n \in \mathbb{Z}$, we define a function $\chi_n : \mathbb{T} \to \mathbb{C}$ by

$$\chi_n(t) = e^{2\pi int} = \cos 2\pi nt + i \sin 2\pi nt,$$

   where we have employed *Euler's formula* for the *trigonometric exponential* functions χ_n. We note that χ_n is well defined on cosets of $\mathbb{Z}$.[110]

  ii. For each $f \in L^1(\mathbb{T})$ or in $L^2(\mathbb{T})$, we define the *Fourier transform* to be a function $\hat{f} : \mathbb{Z} \to \mathbb{C}$ by

$$\hat{f}(n) = \int_{\mathbb{T}} f(t)\overline{\chi_n(t)}\, dl,$$

   where the integral over $\mathbb{T}$ of a periodic function f of period 1 means the integral over any interval of length 1.

 iii. We define the Fourier series $S(f)$ of $f \in L^1(\mathbb{T})$ to be

$$\sum_{n \in \mathbb{Z}} \hat{f}(n)\chi_n,$$

   making no claim regarding the convergence or divergence of the Fourier series in any sense. The Nth *partial sum* of the Fourier series is denoted by

$$S_N(f) = \sum_{n=-N}^{N} \hat{f}(n)\chi_n$$

for each $N \in \mathbb{N}$.

It is important to understand that each of the functions χ_n is a homomorphism of the additive circle group, $\mathbb{R}/\mathbb{Z}$, to the multiplicative group of the complex unit circle.[111] Especially, one must observe that

$$\chi_n(s + t) \equiv \chi_n(s)\chi_n(t).$$

Theorem 10.1.1 *If $f \in L^1(\mathbb{T})$, the nth partial sum S_n of its Fourier series is given by*

$$S_n(x) = \int_{-\frac{1}{2}}^{\frac{1}{2}} f(t)D_n(x - t)\, dt, \tag{10.1}$$

---

[110]The function χ_n is called a *character* because it is continuous and it has the property that $\chi_n(x+y) = \chi_n(x)\chi_n(y)$ for all $x, y \in \mathbb{R}$.
[111]See Exercise 10.5.

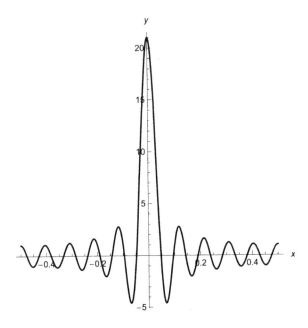

Figure 10.1   Dirichlet kernel D_n for $n = 10$.

where the Dirichlet kernel D_n *is defined by*

$$D_n(x) = \begin{cases} \frac{\sin(2n+1)\pi x}{\sin \pi x} & \text{if } x \notin \mathbb{Z}, \\ 2n + 1 & \text{if } x \in \mathbb{Z}. \end{cases} \tag{10.2}$$

Also, $\displaystyle\int_{-\frac{1}{2}}^{\frac{1}{2}} D_n(x)\, dx = 1,$ *for each* $n \in \mathbb{N}$.[112]

Remark 10.1.2 The Dirichlet kernel does not converge to zero for x bounded away from the origin. It depends for its work on rapid oscillations in sign to produce cancellations, together with most of its integral being nearly 1 over a small interval around the origin. See Figure 10.1.

Proof: We observe that

$$S_n(x) = \sum_{k=-n}^{n} \left(\int_0^1 f(t) \bar{\chi}_k(t)\, dt \right) \chi_k(x) = \sum_{k=-n}^{n} \int_0^1 f(t) \chi_k(x - t)\, dt$$

$$= \int_0^1 f(t) \sum_{k=-n}^{n} \chi_k(x - t)\, dt = \int_{-\frac{1}{2}}^{\frac{1}{2}} f(t) \left(\sum_{k=-n}^{n} e^{2\pi i k(x-t)} \right) dt,$$

---

[112]The theorems in the present section have been adapted from the author's book [20], *Advanced Calculus: An Introduction to Linear Analysis.*

since the integrand has period 1 and can be integrated with the same result on any interval of length 1. It will suffice to prove that the sum inside the integrand is the Dirichlet kernel evaluated at $x - t$. We reason as follows, using Euler's formula and the sum of a geometric series:

$$\sum_{k=-n}^{n} e^{2\pi i k x} = e^{-2\pi i n x} \sum_{k=0}^{2n} e^{2\pi i k x} = e^{-2\pi i n x} \frac{1 - e^{2\pi i (2n+1) x}}{1 - e^{2\pi i x}}$$

$$= \frac{e^{-2\pi i n x} - e^{2\pi i (n+1) x}}{1 - e^{2\pi i x}} = \frac{e^{-i(2n+1)\pi x} - e^{i(2n+1)\pi x}}{e^{-i\pi x} - e^{i\pi x}}$$

$$= \frac{\sin(2n+1)\pi x}{\sin \pi x} = D_n(x),$$

provided that the denominator in the geometric series formula is not zero, which is equivalent to $x \notin \mathbb{Z}$. If $x \in \mathbb{Z}$, then the sum is clearly $2n + 1$.

Finally, since we have shown above that $D_n(x) = \sum_{k=-n}^{n} e^{2\pi i k x}$, it follows readily that $\int_{-\frac{1}{2}}^{\frac{1}{2}} D_n(x)\, dx = 1$ for each $n \in \mathbb{N}$. ∎

The following famous lemma is very useful.

Lemma 10.1.1 (Riemann-Lebesgue Lemma) *If $f \in L^1(\mathbb{T})$, then*

$$\widehat{f}(n) \to 0$$

as $|n| \to \infty$.

The proof of this lemma is left to Exercise 10.1.

For the next lemma, note that differentiation of a function on $\mathbb{T}$ has the same meaning as differentiation of a periodic function of period 1 on $\mathbb{R}$. The notation $C^p(\mathbb{T})$ stands for the vector space of all p-times continuously differentiable periodic functions of period 1, where p can be a natural number or ∞. In particular, if $f \in C^p(\mathbb{T})$, then its pth derivative, $f^{(p)}$, is continuous, provided that $p < \infty$.

Lemma 10.1.2 *Let $f \in C^p(\mathbb{T})$, where $1 \leqslant p < \infty$. Then*

$$\left|\widehat{f}(n)\right| \leqslant \frac{\|f^{(p)}\|_1}{(2\pi |n|)^p}. \tag{10.3}$$

Proof: We begin by applying integration by parts to

$$\widehat{f'}(n) = \int_0^1 f'(x)\overline{\chi_n(x)}\, dx$$

$$= f(x)\overline{\chi_n(x)}\Big|_0^1 - \int_0^1 f(x)\overline{\chi_n}'(x)\, dx$$

$$= 2\pi i n \widehat{f}(n).$$

We iterate this argument a total of p times, obtaining

$$\hat{f}(n) = \frac{\widehat{f^{(p)}}(n)}{(2\pi i n)^p}. \tag{10.4}$$

Finally, we observe that for each function $g \in L^1(\mathbb{T})$, we have

$$|\hat{g}(n)| = \left| \int_0^1 g(x)\overline{\chi_n}(x)\,dx \right|$$

$$\leqslant \int_0^1 |g(x)\overline{\chi_n}(x)|\,dx$$

$$= \|g\|_1.$$

It is clear that $f^{(p)}$ is integrable since it is continuous. ∎

Theorem 10.1.2 *Let $f \in C^p(\mathbb{T})$, where $1 \leqslant p < \infty$. Then the Nth partial sum*

$$S_N = \sum_{n=-N}^{N} \hat{f}(n)\chi_n$$

converges uniformly to f on the real line. Moreover,

$$\|S_N - f\|_\infty \leqslant KN^{\frac{1}{2}-p},$$

for some constant K that is independent of N but is dependent upon f and p.

Proof: It would suffice for the first part of the theorem to give a proof for $p = 1$, but the the first part follows from the inequality that is the second part, and that is what we will prove. Our first step will be to prove that the sequence S_n is Cauchy in the sup-norm and that it converges at the rate claimed. For each $n \in \mathbb{N}$ and for each $m \geqslant n$ we have

$$|S_m(x) - S_n(x)| \leqslant \sum_{|k|>n} \left| \hat{f}(k) \right|$$

$$\overset{(1)}{\leqslant} \left(\sum_{|k|>n} \left| \widehat{f^{(p)}}(k) \right|^2 \right)^{\frac{1}{2}} \left(\sum_{|k|>n} \frac{1}{(2\pi k)^{2p}} \right)^{\frac{1}{2}}$$

$$\overset{(2)}{\leqslant} \left\| f^{(p)} \right\|_2 \left(2\int_n^\infty (2\pi x)^{-2p}\,dx \right)^{\frac{1}{2}}$$

$$= \left\| f^{(p)} \right\|_2 \frac{(2\pi)^{\frac{1}{2}-p}}{\sqrt{\pi(2p-1)}} n^{\frac{1-2p}{2}} \to 0$$

as $n \to \infty$. For inequality (1) we have used Equation (10.4) and the Cauchy-Schwarz inequality for l_2. For inequality (2) we have used both Bessel's inequality and the integral test for infinite series of positive terms. This proves that

$$\|S_m - S_n\|_{\sup} \leqslant Kn^{\frac{1}{2}-p}$$

for a suitable constant, K, that is independent of n. Thus S_n is uniformly convergent to some continuous function ϕ. Letting $m \to \infty$, we see that

$$\|\phi - S_n\|_{\sup} \to 0$$

as $n \to \infty$, and that the convergence takes place at the rate claimed.

It remains to prove that $S_n \to f$ or, in other words, that $\phi = f$. Since uniform convergence is established already, we need prove only pointwise convergence to f. We fix x arbitrarily and observe that

$$
\begin{aligned}
S_n(x) - f(x) &= \int_{-\frac{1}{2}}^{\frac{1}{2}} f(x+y) D_n(y)\, dy - f(x) \int_{-\frac{1}{2}}^{\frac{1}{2}} D_n(y)\, dy \\
&= \int_{-\frac{1}{2}}^{\frac{1}{2}} \frac{f(x+y) - f(x)}{\sin \pi y} \sin \pi(2n+1)y\, dy \\
&= \int_{-\frac{1}{2}}^{\frac{1}{2}} Q(y) \frac{e^{i\pi(2n+1)y} - e^{-i\pi(2n+1)y}}{2i}\, dy,
\end{aligned}
$$

where we have used Euler's formula, and where we define

$$
Q(y) = \begin{cases} \frac{f(x+y)-f(x)}{\sin \pi y} & \text{if } y \neq 0, \\ \frac{f'(x)}{\pi} & \text{if } y = 0. \end{cases}
$$

Next, we define

$$Q_+(y) = Q(y) e^{i\pi y} \text{ and } Q_-(y) = Q(y) e^{-i\pi y}, \tag{10.5}$$

and the reader can check easily that each of these functions is continuous and thus integrable. Finally, we see that

$$S_n(x) - f(x) = \frac{-i}{2}\left(\widehat{Q_+}(-n) - \widehat{Q_-}(n)\right) \to 0$$

as $n \to \infty$ by the Riemann-Lebesgue lemma. ∎

Theorem 10.1.3 *If $f \in L^2(\mathbb{T})$, then*

$$\|S_n(f) - f\|_2 \to 0$$

as $n \to \infty$. Consequently, we have the Plancherel identity:

$$\sum_{-\infty}^{\infty} \left|\hat{f}(n)\right|^2 = \|f\|_2^2. \tag{10.6}$$

Remark 10.1.3 A celebrated theorem of Lennart Carleson [3] established that the Fourier series of any square-integrable Lebesgue measurable function must converge to $f(x)$ pointwise except on a set of Lebesgue measure zero. However, a set of points can have Lebesgue measure zero and still be an uncountably infinite set. There are examples known of continuous functions f for which $S_n(f)$ is actually *divergent* for infinitely many values of x. And there is an example of a Lebesgue integrable function f for which the Fourier series diverges at each point x! The extraordinary pathologies of Fourier series in regard to pointwise convergence, even for continuous functions, make theorems like the one we are about to prove very interesting and useful.

Proof: In Exercise 10.3, the reader will show that $C^1(\mathbb{T})$ is dense in $L^2(\mathbb{T})$. Let $\epsilon > 0$. It follows that if $f \in L^2(\mathbb{T})$, then there exists a function $\phi \in C^1(\mathbb{T})$ such that

$$\|f - \phi\|_2 < \frac{\epsilon}{2}.$$

Since the partial sums $S_N(\phi)$ of the Fourier series of ϕ converge to ϕ uniformly on $\mathbb{T}$, it follows that there exists $N \in \mathbb{N}$ such that $n \geq N$ implies that

$$\|f - S_n(\phi)\|_2 < \epsilon.$$

By Exercise 10.4, the Fourier coefficients of f provide the optimal L^2-approximation to f. Thus

$$\|f - S_n(f)\|_2 \leq \|f - S_n(\phi)\|_2 < \epsilon$$

for all $n \geq N$. This implies the theorem.   ∎

EXERCISES

10.1 Prove the Riemann-Lebesgue lemma, Lemma 10.1.1. (Hint: Emulate Exercise 5.42.)

10.2 Show that the measure with respect to which we define $L^1(\mathbb{T})$, described in Remark 10.1.1, is invariant under translation by any real number.

10.3 Use the following steps to prove that the set $C^\infty(\mathbb{T})$ is dense in $L^2(\mathbb{T})$.

   a) Prove that the family $\mathcal{S}$ of all step functions is dense in $L^2(\mathbb{T})$. (Hint: See Exercise 5.41.)
   b) Prove that each step function can be approximated as accurately as desired in the L^2-norm by means of a continuous function.
   c) Prove that each continuous function on $\mathbb{T}$ can be approximated as accurately as desired by means of a C^∞-function.[113]

10.4 Suppose that f lies in a separable Hilbert space $\mathcal{H}$, and let $\{e_k \mid k \in \mathbb{Z}\}$ be any countable orthonormal set—not necessarily a basis. Let $c_{-n}, \ldots, c_n$ be any $2n + 1$ complex constants. Then

$$\left\| f - \sum_{k=-n}^{n} \hat{f}(k)e_k \right\|_2 \leq \left\| f - \sum_{k=-n}^{n} c_k e_k \right\|_2,$$

---

[113] You can use the Weierstrass polynomial approximation theorem for this. See, for example, [20].

with equality holding if and only if $c_k = \hat{f}(k)$ for each k. (Hint: See Definition 9.4.2. Write the difference inside the right-side norm as the sum of two sums of differences, one of these involving $\left(\hat{f}(k) - c_k\right)e_k$.)

10.5   Denote by $(\mathbb{T}, \cdot)$ the multiplicative group of complex numbers of modulus one to distinguish it from the additive group $(\mathbb{T}, +) = \mathbb{R}/\mathbb{Z}$. We call a function $\chi : \mathbb{T} \to (\mathbb{T}, \cdot)$ a *character* of T if and only if χ is both a continuous map and a homomorphism, meaning that

$$\chi(x + y) = \chi(x)\chi(y),$$

for all x and y. Let $\hat{\mathbb{T}}$ denote the set of all characters of $\mathbb{T}$.

  a) Prove that $\hat{\mathbb{T}}$ is a group under the operation of pointwise multiplication. We will call $\hat{\mathbb{T}}$ the *character group* of $\mathbb{T}$.
  b) Prove that $\chi \in \hat{\mathbb{T}}$ if and only if $\chi = \chi_n$, as defined in Definition 10.1.1, for some $n \in \mathbb{Z}$. Hints: Necessity is the harder part. Let χ be any character of $\mathbb{T}$. Prove that

$$\chi(x)\int_0^t \chi(y)\,dy = \int_x^{x+t} \chi(u)\,du$$

and explain why $\chi \in C^1(\mathbb{T})$. Prove that χ' must be a constant multiple of itself.

10.6   Let $\mathbb{T}_T = (\mathbb{R}, +)/(T\mathbb{Z}, +)$ denote the *nonstandard* additive circle group, which is a circle of perimeter $T > 0$. Use the isomorphism $\tau : \mathbb{R}/\mathbb{Z} \to \mathbb{R}/(T\mathbb{Z})$ defined by $\tau(x) = Tx$ to show that

$$\left\{ e_n = \frac{1}{\sqrt{T}}e^{2\pi i \frac{n}{T}x} \,\middle|\, n \in \mathbb{Z} \right\}$$

is an orthonormal basis for $L^2(\mathbb{T}_T)$ and that the Fourier series of a function f in $C^p(\mathbb{T}_T)$ converges uniformly to f, just as in Theorem 10.1.2. (In the case of $\mathbb{R}/(T\mathbb{Z})$, the interval $[0, T]$ plays the role of a convenient cross section for the quotient group, and a function on such a circle of perimeter T would be a periodic function on $\mathbb{R}$ with period T.)

10.2   CLOSED, INVARIANT SUBSPACES OF $L^2(\mathbb{T})$

In this section, we will classify all the closed, translation-invariant subspaces of $L^2(\mathbb{T})$. We begin with a reinterpretation of Theorem 10.1.3.

Definition 10.2.1 Let K be a set of indices, which may be either finite or infinite. Suppose there is a set of Hilbert spaces $\mathcal{H}_k$, indexed by the set K. The *direct sum*

$$\mathcal{H} = \bigoplus_{k \in K} \mathcal{H}_k$$

of the Hilbert spaces H_k is the set of all *functions* f such that $f(k) \in \mathcal{H}_k$ for each $k \in K$ and such that

$$\sum_{k \in K} \|f(k)\|_k^2 < \infty.$$

We define a Hermitian scalar product on $\mathcal{H}$ by

$$\langle f, g \rangle = \sum_{k \in K} \langle f(k), g(k) \rangle_k,$$

where $\langle \cdot, \cdot \rangle_k$ denotes the scalar product in the Hilbert space $\mathcal{H}_k$.

The reader will verify in Exercise 10.7 that the direct sum, as defined above, is a Hilbert space in its own right.

Theorem 10.1.3 tells us that for each $f \in L^2(\mathbb{T})$, we have

$$f = \sum_{n \in \mathbb{Z}} \hat{f}(n) \chi_n$$

in the sense of L^2-convergence, and that the set $\{\chi_n \mid n \in \mathbb{Z}\}$ is a complete orthonormal basis for $L^2(\mathbb{T})$. We can interpret the latter theorem as providing a decomposition of the Hilbert space $L^2(\mathbb{T})$ into the direct sum of minimal, nontrivial, mutually orthogonal, translation-invariant subspaces:

$$L^2(\mathbb{T}) = \bigoplus_{n \in \mathbb{Z}} \mathbb{C}\chi_n.$$

Each of the spaces $\mathbb{C}\chi_n$ is a one-dimensional Hilbert space consisting of all the complex scalar multiples of a single character function, χ_n, which does lie in $L^2(\mathbb{T})$. The one-dimensionality of each space $\mathbb{C}\chi_n$ guarantees that it has no nontrivial, proper, translation-invariant subspaces.

It is easy to see that if $S \subseteq \mathbb{Z}$, then

$$V = \bigoplus_{n \in S} \mathbb{C}\chi_n \tag{10.7}$$

is a closed, translation-invariant subspace of $L^2(\mathbb{T})$. We will prove that every closed, translation-invariant subspace has this same form. In words, we will prove that each closed, translation-invariant subspace of $L^2(\mathbb{T})$ is the direct sum of some finite or countable collection of the one-dimensional spaces of the form $\mathbb{C}\chi_n$. We begin with a brief introduction to the integration of Hilbert space valued functions.

10.2.1   Integration of Hilbert Space Valued Functions

We begin with a brief digression in order to define the concept of the integral of a Hilbert space valued function. We have seen in Theorem 9.4.2 that the bounded linear functionals on a separable Hilbert space $\mathcal{H}$ correspond uniquely to the points of $\mathcal{H}$ itself via the bijection

$$T_\xi(\eta) = \langle \eta, \xi \rangle$$

for each $\xi \in \mathcal{H}$.

Definition 10.2.2 Suppose $(X, \mathfrak{A}, \mu)$ is a measure space and suppose that ϕ, mapping X to the separable Hilbert space $\mathcal{H}$, has the properties that the function

$$x \rightarrow \langle \eta, \phi(x) \rangle$$

is μ-measurable, for each fixed $\eta \in \mathcal{H}$. Suppose also that the real-valued function $\|\phi(x)\|_{\mathcal{H}}$ lies in $L^1(X, \mathfrak{A}, \mu)$. Then we define the Hilbert space value of the integral of ϕ with respect to μ by the equation

$$\left\langle \eta, \int_X \phi(x)\, d\mu(x) \right\rangle = \int_X \langle \eta, \phi(x) \rangle\, d\mu(x) \tag{10.8}$$

for each $\eta \in \mathcal{H}$.

For this definition to make sense, it is necessary to show that the right side of Equation (10.8) defines a bounded linear functional

$$T_\phi(\eta) = \int_X \langle \eta, \phi(x) \rangle\, d\mu(x),$$

so that the latter integral determines a unique vector in $\mathcal{H}$,[114] which we call the *value* of the Hilbert space valued integral. This is Exercise 10.8. In the following lemma, we prove a useful generalization of the triangle inequality for the Hilbert space norm of a Hilbert space valued integral.

Lemma 10.2.1 (Triangle Inequality) *Suppose that $\phi : X \rightarrow \mathcal{H}$ satisfies the hypotheses in Definition 10.2.2. Then*

$$\left\| \int_X \phi(x)\, d\mu(x) \right\| \leq \int_X \|\phi(x)\|\, d\mu(x),$$

where the norm on each side of the inequality is the Hilbert space norm.

Proof: We reason as follows, using the definition of the norm of a linear functional:

$$\left\| \int_X \phi(x)\, d\mu(x) \right\| = \sup_{\|\eta\|=1} \left| \left\langle \eta, \int_X \phi(x)\, d\mu(x) \right\rangle \right|$$

$$= \sup_{\|\eta\|=1} \left| \int_X \langle \eta, \phi(x) \rangle\, d\mu(x) \right|$$

$$\leq \sup_{\|\eta\|=1} \int_X |\langle \eta, \phi(x) \rangle|\, d\mu(x)$$

$$\leq \int_X \|\phi(x)\|\, d\mu(x).$$

Here we have used both the definition of the Hilbert space valued integral and the Cauchy-Schwarz inequality.   ∎

---

[114]Note that $\mathcal{H}$ is self-dual—a fact that is important here.

10.2.2 Spectrum of a Subset of $L^2(\mathbb{T})$

Definition 10.2.3 Let E be any subset of $L^2(\mathbb{T})$. We define the *spectrum* of E to be the subset of $\widehat{\mathbb{T}}$ given by

$$\widehat{E} = \bigcup_{f \in E} \left\{ n \,\middle|\, \widehat{f}(n) \neq 0 \right\}.$$

Theorem 10.2.1 *A linear subspace $V \subseteq L^2(\mathbb{T})$ is closed and translation invariant if and only if*

$$V = \bigoplus_{n \in \widehat{V}} \mathbb{C}\chi_n.$$

Proof: Sufficiency is established by Exercise 10.7. We will prove necessity. To this end, denote the *convolution*

$$
\begin{aligned}
f * \chi_n(x) &\overset{(i)}{=} \int_{\mathbb{T}} f(x-t)\chi_n(t)\, dt \\
&= \int_{\mathbb{T}} f(u)\chi_n(x-u)\, du \qquad\qquad (10.9) \\
&= \widehat{f}(n)\chi_n(x) \\
&\overset{(ii)}{=} \int_{\mathbb{T}} f_{-t}(x)\chi_n(t)\, dt.
\end{aligned}
$$

Equality (i) expresses the definition of the convolution over the circle group. Equality (ii) rewrites the right side of Equality (i) in a manner that is useful for the present argument. Since the characters of $\mathbb{T}$ are mutually orthogonal, it will suffice to show that for each $f \in V$ and for each $n \in \mathbb{Z} = \widehat{\mathbb{T}}$, we must have $f * \chi_n \in V$.

Let $\epsilon > 0$, and fix V. Observe that the mapping $t \to f_{-t}$, carrying t to the translation of f by $-t$, is a continuous map from $\mathbb{T}$ to $L^2(\mathbb{T})$.[115] This mapping is uniformly continuous since $\mathbb{T}$ is compact. Thus there exists $\delta > 0$ such that

$$|t - t'| < \delta \implies \|f_t - f_{t'}\|_2 < \epsilon.$$

The plan of the proof is to show that the convolution integral in Equation (10.9) is a limit of a finite linear combination of translations and is therefore contained within

[115] See Exercise 10.9.

the closed subspace V. Let $\frac{1}{m} < \delta$. We reason as follows, applying Lemma 10.2.1:

$$\left\| f * \chi_n(x) - \sum_{k=0}^{m-1} f_{-\frac{k}{m}}(x) \int_{\frac{k}{m}}^{\frac{k+1}{m}} \chi_n(t)\, dt \right\|_2$$

$$= \left\| \sum_{k=0}^{m-1} \int_{\frac{k}{m}}^{\frac{k+1}{m}} \left[f_{-t}(x) - f_{-\frac{k}{m}}(x) \right] \chi_n(t)\, dt \right\|_2$$

$$\leqslant \sum_{k=0}^{m-1} \left\| \int_{\frac{k}{m}}^{\frac{k+1}{m}} \left[f_{-t}(x) - f_{-\frac{k}{m}}(x) \right] \chi_n(t)\, dt \right\|_2$$

$$\leqslant \sum_{k=0}^{m-1} \int_{\frac{k}{m}}^{\frac{k+1}{m}} \left\| f_{-t}(x) - f_{-\frac{k}{m}}(x) \right\|_2 |\chi_n(t)|\, dt$$

$$< \sum_{k=0}^{m-1} \epsilon \frac{1}{m} = \epsilon$$

for all sufficiently large values of m. This completes the proof that $f * \chi_n \in V$, since V is closed. Hence V contains the closed linear span of all the characters in its own spectrum. The definition of the spectrum completes the proof.   ∎

Remark 10.2.1 Although we will not prove it in this book, the fact that the irreducible translation-invariant subspaces are one-dimensional is a reflection of the circle group being both abelian and compact. The real line is abelian, but noncompact, so $L^2(\mathbb{R})$ is a direct integral of one-dimensional irreducible, invariant spaces (but not a direct sum of subspaces). Harmonic analysis can be carried out on matrix groups, on Lie groups, and on locally compact Hausdorff topological groups. For groups that are compact but not abelian, minimal translation-invariant subspaces will be finite-dimensional, but they need not be one-dimensional. For groups G that are neither compact nor abelian, minimal translation-invariant spaces in the direct integral decomposition of $L^2(G)$ can be infinite-dimensional.[116] An interesting example of such an infinite-dimensional, irreducible space under the action of a nonabelian, noncompact group is given in Section 10.5, where the group under consideration is the Heisenberg group. For a well written elementary introduction to topics mentioned in this remark, see the book [19] by Pukanszky.

EXERCISES

10.7 Prove that Definition 10.2.1 makes the direct sum of an indexed family of Hilbert spaces into a Hilbert space.

10.8 Prove that Equation (10.8) determines the integral uniquely as a vector in $\mathcal{H}$ by identifying its action as a point in the dual space.

---

[116]Such groups play an important role in quantum mechanics.

10.9 Let $f \in L^2(\mathbb{T})$. Prove that the mapping $t \to f_{-t}$ is a continuous map from $\mathbb{T}$ to $L^2(\mathbb{T})$. (Hint: See Exercises 9.15 and 5.43.)

10.3 SCHWARTZ FUNCTIONS: FOURIER TRANSFORM AND INVERSION

The study of the representation of functions $f \in L^2(\mathbb{T})$ by Fourier series was facilitated by the study of $C^\infty(\mathbb{T})$ first. For the real line itself, there is a special role played by the *Schwartz* functions, which we will define below. First, we define the *characters* of $\mathbb{R}$ and the Fourier transform.

Definition 10.3.1 Let $f \in L^1(\mathbb{R})$ and let $\gamma \in \mathbb{R}$. We make the following definitions.

  i. For each $\gamma \in \mathbb{R}$, we define the *character* $\chi_\gamma : \mathbb{R} \to (\mathbb{T}, \cdot)$ by

$$\chi_\gamma(x) = e^{2\pi i \gamma x}.$$

Denote the set of all characters of $\mathbb{R}$ by

$$\widehat{\mathbb{R}} = \{\chi_\gamma \mid \gamma \in \mathbb{R}\}.$$

  ii. Define the *Fourier transform* of the function f to be $\widehat{f} : \widehat{\mathbb{R}} \to \mathbb{C}$, given by

$$\widehat{f}(\gamma) = \int_{\mathbb{R}} f(x)\overline{\chi_\gamma(x)} \, dx.$$

  iii. Define the *inverse Fourier transform* of $g \in L^1(\mathbb{R})$ to be

$$\check{g}(x) = \int_{\mathbb{R}} g(\gamma)\chi_\gamma(x) \, d\gamma.$$

In Exercise 5.47, the reader will have proven that the Fourier transform of each function $f \in L^1(\mathbb{R})$ is continuous and that it vanishes at $\pm\infty$.[117] The reader should check the simple extension that the inverse Fourier transform of an L^1-function has the same properties: In fact, $\check{g}(x) = \widehat{g}(-x)$.

In Exercise 10.10, the reader will show that $\widehat{\mathbb{R}}$ is the abelian group of all continuous homomorphisms of $(\mathbb{R}, +) \to (\mathbb{T}, \cdot)$. We leave it as an informal and very easy exercise for the reader to prove that the integral defining the Fourier transform of each integrable function f on $\mathbb{R}$ exists.

The wish would be to show that

$$\check{\widehat{f}} = f,$$

---

[117]This is the Riemann-Lebesgue lemma.

for each $f \in L^1(\mathbb{R})$. However, it is not hard to give examples of integrable functions f for which $\hat{f}$ is not integrable, leaving its inverse Fourier transform undefined. (See Exercise 10.11.) We are ready to define the family of all Schwartz [118] functions on $\mathbb{R}$. These functions will not share the difficulty just described for $L^1(\mathbb{R})$.

Definition 10.3.2 We define the set of all Schwartz functions on the real line to be

$$ \mathcal{S}(\mathbb{R}) = \left\{ f \in \mathcal{C}^{\infty}(\mathbb{R}) \,\middle|\, \left(1 + x^2\right)^k f^{(n)}(x) \to 0 \text{ as } |x| \to \infty \;\forall k, n \in \mathbb{N} \cup \{0\} \right\}. $$

In words, Schwartz functions can be described as being those smooth functions on the line that vanish rapidly at plus and minus infinity, as does every derivative of every finite order n. The role of the index k is to ensure that the Fourier transform of each derivative of a Schwartz function vanishes more rapidly at plus and minus infinity than the reciprocal of any polynomial. The reader should check easily that every Schwartz function is integrable and that every derivative of a Schwartz function is Schwartz.

Theorem 10.3.1 *The Fourier transform,* $\hat{} : f \to \hat{f}$ *for each* $f \in L^1(\mathbb{R})$, *has the following properties when restricted to* $\mathcal{S}(\mathbb{R})$:

   i. *The Fourier transform is a* bijection *of* $\mathcal{S}(\mathbb{R})$ *onto itself.*

   ii. $\overset{\smile}{\hat{f}} = f$ *for each* $f \in \mathcal{S}(\mathbb{R})$.

   iii. *The Fourier transform is an* L^2-*isometry of* $\mathcal{S}(\mathbb{R})$, *meaning that* $\|f\|_2 = \left\|\hat{f}\right\|_2$ *for each* $f \in \mathcal{S}(\mathbb{R})$. [119]

Proof:

   i. We can produce a variant of Lemma 10.1.2, applying integration by parts to show that for Schwartz functions on the real line we have

$$ \left|\hat{f}(\gamma)\right| \leq \frac{\left\|f^{(p)}\right\|_1}{|2\pi\gamma|^p}, \tag{10.10} $$

for all $\gamma \neq 0$. We see easily that

$$ \left|\hat{f}(0)\right| \leq \int_{\mathbb{R}} |f| \, dl = \|f\|_1. $$

From this it follows readily that the Fourier transform $\hat{f}$ of a Schwartz function f must be rapidly decreasing. We need to prove that $\hat{f}$ is differentiable and that it is in $\mathcal{C}^{\infty}(\mathbb{R})$ as well. We will apply the Lebesgue Dominated Convergence

---

[118]Schwartz functions are named for Laurent Schwartz, who lived later than the inventors of the Cauchy-Schwarz inequality. The difference in spelling is not accidental.
[119]Clearly, each Schwartz function on the line is also square integrable.

theorem, together with the mean value inequality for complex-valued functions of a real variable.[120] Select an arbitrary sequence $h_n \to 0$. It is necessary to prove that

$$\left(\hat{f}\right)'(\gamma) = \lim_{n \to \infty} \int_{\mathbb{R}} f(x) \frac{\overline{\chi}_{\gamma+h_n}(x) - \overline{\chi}_\gamma(x)}{h_n} \, dx \qquad (10.11)$$
$$= -2\pi i \widehat{(xf)}(\gamma)$$

by showing that this limit exists and is independent of the choice of $h_n \to 0$. We leave the verification of this to the reader in Exercise 10.15. Applying Equation (10.11), and using the fact that $xf \in \mathcal{S}(\mathbb{R})$, we see that $\left(\hat{f}\right)'(\gamma) \to 0$ fast as $|\gamma| \to \infty$. By mathematical induction, we establish that the Fourier transform maps $\mathcal{S}(\mathbb{R})$ into itself.

We leave the proof that the Fourier transform is surjective until the end of this proof. *For the second and third parts of the proof, we will treat first the case of functions* $f \in \mathcal{C}_c^\infty(\mathbb{R})$, which is the linear subspace of $\mathcal{S}(\mathbb{R})$ consisting of all infinitely differentiable functions with compact support.

ii. If $f \in \mathcal{C}_c^\infty(\mathbb{R})$, then there exists a sufficiently big real number $T > 0$ such that the interval $\left[-\frac{T}{2}, \frac{T}{2}\right)$ contains the support of f. By Exercise 10.6, we claim that

$$f(x) \overset{(a)}{=} \sum_{n \in \mathbb{Z}} \int_{-\frac{T}{2}}^{\frac{T}{2}} f(y) \frac{e^{-2\pi i \frac{n}{T} y}}{\sqrt{T}} \, dy \, \frac{e^{2\pi i \frac{n}{T} x}}{\sqrt{T}}$$
$$\overset{(b)}{=} \sum_{n \in \mathbb{Z}} \hat{f}\left(\frac{n}{T}\right) \frac{e^{2\pi i \frac{n}{T} x}}{T}$$
$$= \sum_{n \in \mathbb{Z}} \int_{\mathbb{R}} \hat{f}\left(\frac{n}{T}\right) e^{2\pi i \frac{n}{T} x} 1_{\left[\frac{n}{T}, \frac{n+1}{T}\right)}(\gamma) \, d\gamma$$
$$\overset{(c)}{\to} \int_{\mathbb{R}} \hat{f}(\gamma) e^{2\pi i \gamma x} \, d\gamma = \overset{\vee}{\hat{f}}(x)$$

as $T \to \infty$. In Equality (a) we use a Hilbert space Fourier series using an orthonormal basis, and in Equality (b) we use the form of the Fourier transform for the real line (rather than the circle) on the right-hand side. The convergence claim (c) is what we will prove now for $f \in \mathcal{C}_c^\infty(\mathbb{R})$. The details are as follows.

For a suitable constant c, we have

$$\left|\hat{f}(\gamma)\right| \leq \frac{c}{1 + \gamma^2}$$

---

[120]This inequality is a special case of the Euclidean space Mean Value Theorem, interpreting $f : \mathbb{R} \to \mathbb{C}$ as though it were $\mathbb{R}^2$-valued. See [20], for example.

for all γ because f is Schwartz. Assume without loss of generality that $T > 1$, and define the function $h \in L^1(\mathbb{R})$ by

$$h(\gamma) = \sup_{\gamma-1 \leqslant t < \gamma} \frac{c}{1 + t^2}.$$

Then we have

$$\left| \hat{f}\left(\frac{n}{T}\right) e^{2\pi i \frac{n}{T} x} 1_{\left[\frac{n}{T}, \frac{n+1}{T}\right)}(\gamma) \right| \leqslant h(\gamma),$$

and we note that the supports of the functions of γ on the left-hand side are mutually disjoint. Also,

$$\hat{f}\left(\frac{n}{T}\right) e^{2\pi i \frac{n}{T} x} 1_{\left[\frac{n}{T}, \frac{n+1}{T}\right)}(\gamma) \to \hat{f}(\gamma) e^{2\pi i \gamma x},$$

pointwise on $\mathbb{R}$, as $T \to \infty$. Now the Lebesgue Dominated Convergence theorem tells us that

$$f(x) = \int_{\mathbb{R}} \hat{f}(\gamma) e^{2\pi i \gamma x} \, d\gamma = \overset{\smile}{\hat{f}}(x).$$

iii. Again for this part, we assume for now that $f \in C_c^{\infty}(\mathbb{R})$. Just as in the preceding part, we have for all sufficiently big $T > 0$ that

$$\|f\|_2^2 = \int_{-\frac{T}{2}}^{\frac{T}{2}} |f(x)|^2 \, dx$$

$$= \sum_{n \in \mathbb{Z}} \frac{1}{T} \left| \hat{f}\left(\frac{n}{T}\right) \right|^2 \to \int_{\mathbb{R}} \left| \hat{f}(\gamma) \right|^2 \, d\gamma$$

as $T \to \infty$. Hence $\|f\|_2 = \left\| \hat{f} \right\|_2$, making the Fourier transform an isometric injection of $C_c^{\infty}(\mathbb{R})$ into $\mathcal{S}(\mathbb{R})$, with respect to the L^2-norm.

It remains to show that the Fourier transform maps $\mathcal{S}(\mathbb{R})$ isometrically onto itself and that Fourier inversion works for all Schwartz functions. To these ends, we let f be an arbitrary Schwartz function. We would like to approximate f with a sequence of compactly supported smooth functions. Let $\phi \in C_c^{\infty}(\mathbb{R})$ such that $\phi(x) \equiv 1$ on $\left[-\frac{1}{2}, \frac{1}{2}\right]$, with $\|\phi\|_{\infty} = 1$, and the support of ϕ contained in $[-1, 1]$.[121] We define our approximating sequence as follows:

$$f_n(x) = f(x)\phi\left(\frac{x}{n}\right). \tag{10.12}$$

Observe that

$$\left\| \hat{f} - \widehat{f_n} \right\|_{\infty} \leqslant \|f - f_n\|_1$$

$$\leqslant \int_{|x| \geqslant \frac{n}{2}} |f(x)| \, dx \to 0$$

---

[121] The existence of such smooth functions ϕ can be found in many advanced calculus texts, such as [20].

as $n \to \infty$. Now we make a two-part estimation on

$$\left| \widehat{f_n}(\gamma) \right| \leq \left\| \widehat{f_n} \right\|_{\infty} \leq \|f_n\|_1 \leq \|f\|_1$$

for all γ, and in particular for $|\gamma| \leq 1$. But for $|\gamma| > 1$, we make the following estimates:

$$\left| \widehat{f_n}(\gamma) \right| \leq \frac{1}{4\pi^2\gamma^2} \|f_n''\|_1$$

$$\leq \frac{1}{4\pi^2\gamma^2} \left\| f''\|\phi\|_{\infty} + \frac{2}{n} f'\|\phi'\|_{\infty} + \frac{1}{n^2} f\|\phi''\|_{\infty} \right\|_1$$

$$\leq \frac{1}{4\pi^2\gamma^2} \left[\|f''\|_1\|\phi\|_{\infty} + 2\|f'\|_1\|\phi'\|_{\infty} + \|f\|_1\|\phi''\|_{\infty} \right] = \frac{C}{\gamma^2},$$

with C independent of n. Hence the sequence $\left| \widehat{f_n} \right|$ is dominated in the sense of the Lebesgue Dominated Convergence theorem, with the same being true of $\left| \widehat{f_n} \right|^2$.

It follows that

$$f(x) = \lim_{n \to \infty} f_n(x) = \lim_{n \to \infty} \int_{\mathbb{R}} \widehat{f_n}(\gamma) e^{2\pi i \gamma x} \, d\gamma$$

$$\overset{\text{(LDC)}}{=} \int_{\mathbb{R}} \hat{f}(\gamma) e^{2\pi i \gamma x} \, d\gamma = \check{\hat{f}}(x).$$

Also,

$$\left\| \hat{f} \right\|_2^2 = \int_{\mathbb{R}} \left| \hat{f}(\gamma) \right|^2 d\gamma = \int_{\mathbb{R}} \lim_{n \to \infty} \left| \widehat{f_n}(\gamma) \right|^2 d\gamma$$

$$\overset{\text{(LDC)}}{=} \lim_{n \to \infty} \int_{\mathbb{R}} \left| \widehat{f_n}(\gamma) \right|^2 d\gamma$$

$$= \lim_{n \to \infty} \int_{\mathbb{R}} |f_n(x)|^2 \, dx$$

$$= \int_{\mathbb{R}} |f(x)|^2 \, dx = \|f\|_2^2.$$

Hence the Fourier transform is an isometric injection of $\mathcal{S}(\mathbb{R})$ into itself, and we can see that this transformation is surjective by Exercise 10.13.b. ∎

EXERCISES

10.10   Prove that the set $\hat{\mathbb{R}}$ is an abelian group under the operation of pointwise multiplication, and that every continuous homomorphism of the additive group $(\mathbb{R}, +)$ to the multiplicative group $(\mathbb{T}, \cdot)$ of the complex unit circle is an element of $\hat{\mathbb{R}}$. (Hint: If χ is any nontrivial homomorphism of of $(\mathbb{R}, +) \to (\mathbb{T}, \cdot)$, then it has a discrete *kernel*, $\ker(\chi) = \beta\mathbb{Z}$ for some positive real number β. Now consider the *nonstandard* circle, $\mathbb{R}/\beta\mathbb{Z}$.)

10.11 Find and justify an example of a function $f \in L^1(\mathbb{R})$ for which $\hat{f} \notin L^1(\mathbb{R})$.

10.12 Let p be any polynomial, $c > 0$, and $x_0 \in \mathbb{R}$. Show that

$$f(x) = p(x)e^{-c(x-x_0)^2}$$

is a Schwartz function.

10.13 Define the function $\delta_c : \mathbb{R} \to \mathbb{R}$ by $\delta_c x = cx$ for each $c \in \mathbb{R}$. Prove that *for each Schwartz function* $f \in \mathcal{S}(\mathbb{R})$ we have the following identities:

a) $\overset{\smile}{\hat{f}} = \hat{f} \circ \delta_{-1}$.

b) $\overset{\widehat{\widehat{\widehat{\widehat{f}}}}}{} = f$. This is sometimes expressed with the operator equation

$$\widehat{}^4 = I,$$

the identity operator. In words, the fourth power of the Fourier transform is the identity operator. (This exercise may prompt the reader to search for eigenfunctions of the Fourier transform operator, corresponding to the four complex roots of unity. In fact the *Hermite* functions, which are the functions

$$h_n(x) = e^{\pi x^2} \frac{d^n}{dx^n} e^{-2\pi x^2}$$

up to a constant factor, are eigenfunctions. The reader can learn more about this in [5], in which it is shown that h_n is an eigenfunction for the eigenvalue $(-i)^n$.)

10.14 If f and g are in $\mathcal{S}(\mathbb{R})$, show that

$$\left(\widehat{\hat{f}\hat{g}}\right) = f * g.$$

(See Exercise 6.7.b.)

10.15 Prove Equation (10.11) for every function $f \in \mathcal{S}(\mathbb{R})$.

10.4 CLOSED, INVARIANT SUBSPACES OF $L^2(\mathbb{R})$

We will begin by defining the Fourier transform of each $f \in L^2(\mathbb{R})$ by means of Schwartz approximations. Then we will use the L^2-Fourier transform to classify all the closed, translation-invariant subspaces of $L^2(\mathbb{R})$.

10.4.1 The Fourier Transform in $L^2(\mathbb{R})$

We caution the reader that we *cannot* define

$$\hat{f}(\gamma) = \int_{\mathbb{R}} f(x)e^{-2\pi i \gamma x}\, dx,$$

since $L^2(\mathbb{R}) \not\subseteq L^1(\mathbb{R})$, and thus the integral need not exist. This deficiency can be remedied, thanks to the density of $S(\mathbb{R})$ in $L^2(\mathbb{R})$.

Lemma 10.4.1 *The vector space $S(\mathbb{R})$ of Schwartz functions is dense in the Hilbert space in $L^2(\mathbb{R})$.*

Proof: It will suffice to prove that C_c^∞ is dense in $L^2(\mathbb{R})$. We have shown, in Exercise 5.41, that the space S of step functions [122] is dense in $L^1(\mathbb{R})$. This implies that S is dense in $L^2(\mathbb{R})$ as well, since each integral of a nonnegative function is the supremum of the integrals of its truncations in both domain and range. Very much as we did for Equation (10.12), in the proof of Theorem 10.3.1, we can construct a $C_c^\infty(\mathbb{R})$-function ϕ that comes as close as we like in L^2-norm to any given indicator function of an interval, and hence as close as we like to any step function f. The method is to provide a C^∞-interpolation between any two arbitrary heights, a and b, over a width in the domain that is positive but as small as desired. [123] In this manner, a square integrable step function can be approximated as closely as desired by means of a C_c^∞-function, completing the proof of the lemma. ∎

If we take a sequence of Schwartz functions f_n such that $\|f_n - f\|_2 \to 0$, then the fact that f_n is Cauchy in the L^2-norm implies that the sequence $\widehat{f_n}$ is also L^2-Cauchy, thanks to Theorem 10.3.1. We make the following definition, which will require justification.

Definition 10.4.1 For $f \in L^2(\mathbb{R})$, we define $\hat{f}$ to be the L^2-limit of $\widehat{f_n}$ for each sequence of Schwartz functions $f_n \to f$ in the L^2-norm.

It should be noted that we have not defined $\hat{f}$ pointwise as a function of $\gamma \in \mathbb{R}$. Rather, we have defined the Fourier transform as an L^2-equivalence class by invoking the completeness of L^2. We must show that this definition is independent of the choice of the sequence of Schwartz functions f_n. Also, we have at this point two different definitions of the Fourier transform for those functions that are in $L^1(\mathbb{R}) \cap L^2(\mathbb{R})$, and these must be shown to agree.

Lemma 10.4.2 *If f_n and g_n are any two sequences of Schwartz functions converging in the L^2-norm to f, then*

$$\lim_{n \to \infty} \widehat{f_n} = \lim_{n \to \infty} \widehat{g_n}$$

as elements of $L^2(\mathbb{R})$.

Proof: Applying Theorem 10.3.1 again, we see that

$$\left\| \widehat{f_n} - \widehat{g_n} \right\|_2 = \|\widehat{(f_n - g_n)}\|_2 = \|f_n - g_n\|_2$$
$$\leqslant \|f_n - f\|_2 + \|f - g_n\|_2 \to 0.$$

---

[122] The reader should take care not to confuse the space S of step functions with the space $S(\mathbb{R})$ of Schwartz functions.

[123] See Exercise 5.62 in [20] for the details of this useful technique.

■

Corollary 10.4.1 *The Fourier transform is a linear isometry of $L^2(\mathbb{R})$ onto itself. Moreover, the inverse Fourier transform, denoted by $\smallsmile$, is a well-defined isometric surjection of $L^2(\mathbb{R})$. Both transforms preserve the Hermitian scalar product, and this is called* Parseval's Identity *for $L^2(\mathbb{R})$.*

Proof: If we have Schwartz functions $f_n \to f$ and $g_n \to g$, then $af_n + g_n \to af + g$, and

$$(af + g)^\wedge = \lim_{n \to \infty} (af_n + g_n)^\wedge$$
$$= \lim_{n \to \infty} \left(a\widehat{f_n} + \widehat{g_n} \right) = a\widehat{f} + \widehat{g},$$

proving linearity. The Fourier transform is an L^2-isometry because

$$\|f\|_2 = \lim_{n \to \infty} \|f_n\|_2$$
$$= \lim_{n \to \infty} \left\| \widehat{f_n} \right\|_2 = \left\| \widehat{f} \right\|_2.$$

To see that the Fourier transform is a surjection from L^2 onto itself, let $f \in L^2(\mathbb{R})$ be arbitrary. We pick Schwartz functions $f_n \to f$, and we invoke Exercise 10.13.b, which tells us that

$$f = \lim_n f_n = \lim_n \widehat{\widehat{\widehat{\widehat{f_n}}}} = \widehat{\widehat{\widehat{\widehat{f}}}}.$$

Thus f lies in the range of the Fourier transform. This shows also also that the fourth power of the Fourier transform is equal to the identity operator on $L^2(\mathbb{R})$, just as it is on $S(\mathbb{R})$. Moreover, the third power of the Fourier transform must be the inverse Fourier transform: $\breve{g} = \widehat{\widehat{\widehat{g}}}$ for all $g \in L^2(\mathbb{R})$.

The final conclusion follows from the Parallelogram Law, which determines the Hermitian scalar product of a complex inner product space in terms of its associated norm. ■

We should note that if $f \in L^1(\mathbb{R}) \cap L^2(\mathbb{R})$, then we have defined $\widehat{f}$ in two distinct ways: first as an integral and then as a limit of transforms of Schwartz functions. The following lemma establishes that these two definitions for the Fourier transform of such a function coincide.

Lemma 10.4.3 *For each $f \in L^1(\mathbb{R}) \cap L^2(\mathbb{R})$, the L^2-Fourier transform of f is equal almost everywhere to $\int_{\mathbb{R}} f(x)e^{-2\pi i\gamma x}\, dx$.*

Proof: Let $f_N = f 1_{[-N,N]}$ for each $N \in \mathbb{N}$. We see that f_N converges in the norms of both L^1 and L^2 to f. Denote

$$\widehat{f_N}(\gamma) = \int_{\mathbb{R}} f_N(x)e^{-2\pi i\gamma x}\, dx,$$

the L^1-Fourier transform of f_N. Lebesgue Dominated Convergence tells us that $\widehat{f_N} \rightarrow \hat{f}$, the L^1-Fourier transform of f. From Lemma 5.5.1 concerning the pointwise convergence of rapidly L^1-Cauchy sequences, we know that there is a *subsequence* $\widehat{f_{N_k}}$ that converges pointwise almost everywhere to the L^1-transform of f. However, since f_{N_k} converges also in the L^2-norm to f, the L^2-transforms $\widehat{f_{N_k}}$ converge in L^2-norm to $\hat{f}$. Passing to a suitable subsequence again (as in the proof of completeness of L^p) would ensure pointwise convergence almost everywhere to a function in the L^2-equivalence class of the L^2-Fourier transform of f.

Thus it would suffice to know that for the compactly supported function f_N, the L^1- and L^2-Fourier transforms coincide. To this end, pick a sequence φ_n in $C^\infty[-N, N]$ that converges L^2 to f_N. Note that

$$\langle \phi_n, \chi_\gamma \rangle \rightarrow \langle f_N, \chi_\gamma \rangle$$

and that the latter Hermitian scalar product is the L^1-Fourier transform of f_N, expressed in terms of the scalar product for $L^2[-N, N]$. Here we benefit from the fact that the characters $\chi_\gamma \in L^2[-N, N]$, though they do not belong to $L^2(\mathbb{R})$. By passing implicitly to a subsequence in n if needed, we can assume without loss of generality that

$$\widehat{\phi_n}(\gamma) \rightarrow \widehat{f_N}(\gamma)$$

for almost all γ. But since ϕ_n converges L^2 to f_N, we have $\widehat{\phi_n}$ converging in the L^2-norm to the L^2-transform $\widehat{f_N}$. Passing again, as needed, to a suitable subsequence, we get pointwise convergence almost everywhere. Thus the two concepts of Fourier transform for f_N agree almost everywhere, and the proof is complete. ∎

10.4.2 Translation-Invariant Subspaces of $L^2(\mathbb{R})$

If H is a nontrivial closed, proper subspace of $L^2(\mathbb{R})$, then Exercise 9.19 ensures that H has an orthogonal complement $H^\perp$ such that

$$L^2(\mathbb{R}) = H \oplus H^\perp.$$

If H is a translation-invariant subspace, then the reader will show in Exercise 10.16 that this implies the translation invariance of $H^\perp$ as well.

Lemma 10.4.4 *Let E be any Lebesgue measurable subset of the real line, $\mathbb{R}$, and let*

$$H = \left\{ f \in L^2(\mathbb{R}) \,\middle|\, \hat{f}(\gamma) = 0 \text{ ae on } E^c = \mathbb{R}\backslash E \right\}. \tag{10.13}$$

Then H is a closed, translation-invariant subspace of $L^2(\mathbb{R})$, and we call E the spectrum, $\widehat{H}$, of the closed, translation-invariant subspace H.

Another measurable set, E', would have the property that $\widehat{H} = E'$ if and only if $l(E \triangle E') = 0$, so that $E \sim E'$. That is, the spectrum of a closed, translation-invariant subspace of $L^2(\mathbb{R})$ is determined up to a null set.

Proof: Recall that the Fourier transform is an isometry of $L^2(\mathbb{R})$. It follows that if $f_n \to f$ is any Cauchy sequence in H, then $\widehat{f_n}$ is Cauchy as well, and $\widehat{f_n} \to \hat{f}$ in L^2-norm. Some subsequence of $\widehat{f_n}$ is pointwise convergent almost everywhere. This implies that $\hat{f}(\gamma) = 0$ almost everywhere on E^c, so that $f \in H$, which is thus shown to be closed.

We claim that H is translation-invariant as well. Let $f \in H$ and take any sequence of Schwartz functions $f_n \to f$ in the L^2-norm. For each real number a denote $f_a(x) = f(x+a)$, the translation of f by a. Then $\|(f_n)_a - f_a\|_2 \to 0$ as well. But

$$\widehat{(f_n)_a}(\gamma) = \chi_\gamma(a)\widehat{f_n}(\gamma).$$

It follows again by a subsequence argument that $\hat{f}_a(\gamma) = 0$ almost everywhere on E^c, proving that $f_a \in H$. ∎

The next theorem asserts that every closed, translation-invariant subspace of $L^2(\mathbb{R})$ is determined by its spectrum, as in Equation (10.13). This will yield a bijection between the family of all closed, translation-invariant subspaces of $L^2(\mathbb{R})$ and the metric space of measurable subsets of $\mathbb{R}$, in which two sets are identified if the measure of their symmetric difference is zero.

Theorem 10.4.1 *For each closed, translation-invariant subspace H in $L^2(\mathbb{R})$, there is a measurable set E, determined up to a null set, which serves as the spectrum, $\hat{H}$, of H as in Equation (10.13).*

Proof: Let H be any closed, proper, nontrivial, translation-invariant subspace of $L^2(\mathbb{R})$. Then

$$L^2(\mathbb{R}) = H \oplus H^\perp$$

by Exercise 9.19. For each vector $f \in L^2(\mathbb{R})$, there exists a unique decomposition of the form $f = Pf + P^\perp f$, where $Pf \in H$ and $P^\perp f \in H^\perp$. The mappings P and $P^\perp$ are called the *orthogonal projections* onto H and $H^\perp$, respectively. Since $L^2(\mathbb{R})$ is separable, there exists a countable dense subset f_n, and the reader will show in Exercise 10.17 that the set $\{Pf_n \mid n \in \mathbb{N}\}$ is dense in H. Thus H is separable, and it has an orthonormal basis $\{e_n \mid n \in \mathbb{N}\}$. Define a measurable set

$$E = \bigcup_{n \in \mathbb{N}} \{\gamma \in \mathbb{R} \mid \widehat{e_n}(\gamma) \neq 0\}.$$

Note that for each $f \in H$,

$$f = \sum_{n \in \mathbb{N}} \langle f, e_n \rangle e_n,$$

an L^2-convergent sum, so that $\hat{f}\big|_{E^c} = 0$ almost everywhere.

We need to show that if $f \in L^2(\mathbb{R})$ is chosen subject only to the requirement that $\hat{f} = 0$ almost everywhere on E^c, then $f \in H$, implying that E is the spectrum, $\hat{H}$, of H. To this end, let

$$g = f - \sum_{n \in \mathbb{N}} \langle f, e_n \rangle e_n,$$

and it will suffice to prove that $g = 0$. We know at this point that $\hat{g} = 0$ on E^c, so it will suffice to prove that $\hat{g}\big|_E = 0$ as well. As we have observed earlier, for each translation of an L^2 function ϕ by y, $\widehat{\phi_y}(\gamma) = \chi_\gamma(y)\,\hat{\phi}(\gamma)$ almost everywhere. So take $\phi \in H$ and recall that, by its definition, $g \in H^\perp$. Thus, by Parseval's Identity (Corollary 10.4.1) for $L^2(\mathbb{R})$,[124]

$$0 = \langle g, \phi_y \rangle = \left\langle \hat{g}, \chi_\gamma(y)\,\hat{\phi} \right\rangle$$
$$= \int_{\mathbb{R}} \hat{g}(\gamma)\,\overline{\hat{\phi}(\gamma)}\, e^{-2\pi i \gamma y}\, d\gamma$$
$$= \left(\hat{g}\,\overline{\hat{\phi}} \right)^{\widehat{}}(y)$$

for all $y \in \mathbb{R}$. Thus $\hat{g}\,\overline{\hat{\phi}} = 0$ almost everywhere, including on E. Since ϕ was arbitrary, and could have been any of the functions e_n for example, it follows that $\hat{g} = 0$ even on E, so that $g = 0$. ∎

10.4.3 The Fourier Transform and Direct Integrals

In Theorem 10.2.1, we showed how to decompose $L^2(\mathbb{T})$ into the direct sum of one-dimensional, translation-invariant subspaces of $L^2(\mathbb{T})$. Each one-dimensional translation-invariant subspace of $L^2(\mathbb{T})$ is called also an *irreducible* translation-invariant subspace, because it has no proper, closed, nontrivial, translation-invariant subspaces. Indeed, each of the one-dimensional spaces has no nontrivial, proper subspaces at all. We will see that the direct sum of Hilbert spaces, as defined in Definition 10.2.1, is a special case of the concept of a direct integral of Hilbert spaces.

Definition 10.4.2 Let $(A, \mathfrak{A}, \mu)$ be a measure space. Suppose there is a set of Hilbert spaces $\mathcal{H}_\alpha$ indexed by the set A. The *direct integral*

$$\mathcal{H} = \int_A^\oplus \mathcal{H}_\alpha\, d\mu$$

of the Hilbert spaces H_α is the set of all *functions* f such that $f(\alpha) \in \mathcal{H}_\alpha$ for each $\alpha \in A$ and such that

$$\int_A \|f(\alpha)\|_\alpha^2\, d\mu < \infty.$$

We define a Hermitian scalar product on $\mathcal{H}$ by

$$\langle f, g \rangle = \int_A \langle f(\alpha), g(\alpha) \rangle_\alpha\, d\mu.$$

---

[124]Here we use the Fourier transform according to Definition 10.4.1—not in the sense of an abstract Fourier transform in any separable Hilbert space with respect to a specified orthonormal basis.

where $\langle \cdot, \cdot \rangle_\alpha$ denotes the Hermitian scalar product in the Hilbert space $\mathcal{H}_\alpha$, and where $\| \cdot \|_\alpha$ denotes the corresponding Hilbert space norm.

The reader will show in Exercise 10.18 that the direct integral is itself a Hilbert space. Moreover, the direct sum is a special case of the direct integral, in which the measure μ is counting measure on a countable space. Thus we see that the Fourier transform provides a decomposition of $L^2(\mathbb{T})$ as a countable, discrete direct integral of irreducible, closed, translation-invariant *subspaces* of $L^2(\mathbb{T})$ itself.

The situation is different for $L^2(\mathbb{R})$. Here the Fourier transform provides an isomorphism of Hilbert spaces between $L^2(\mathbb{R})$ and the direct integral over the real line of one-dimensional, irreducible, translation-invariant spaces $H_\alpha = \mathbb{C}\chi_\alpha$. Thus

$$L^2(\mathbb{R}) \cong \int_{\mathbb{R}}^{\oplus} H_\alpha \, d\alpha.$$

However, it is important to note that the spaces H_α are *not* subspaces of $L^2(\mathbb{R})$, since each nonzero function in H_α has constant modulus. Thus the Fourier transform in $L^2(\mathbb{R})$ leads one to a more abstract version of harmonic analysis in which the original space, $L^2(\mathbb{R})$, is analyzed in terms of irreducible, translation-invariant spaces that exist only externally to that original space.

EXERCISES

10.16   If H is any closed, translation-invariant subspace of $L^2(\mathbb{R})$, then $H^\perp$ is also closed and translation-invariant. (Hint: Use the fact that Lebesgue measure is translation-invariant.

10.17   Show that each closed vector subspace of separable Hilbert space is separable.

10.18

    a) Prove that the direct integral of a Hilbert spaces, as in Definition 10.4.2, is itself a Hilbert space with respect to the scalar product from that definition.

    b) Show that the direct sum of Hilbert spaces in Definition 10.2.1 is a special case of the direct integral of Hilbert spaces.

10.5   IRREDUCIBILITY OF $L^2(\mathbb{R})$ UNDER TRANSLATIONS AND ROTATIONS

It is a consequence of Theorem 10.4.1 that every nontrivial, closed, proper, translation-invariant subspace of $L^2(\mathbb{R})$ has nontrivial closed, proper, translation-invariant subspaces. Here we will show that if we act upon $L^2(\mathbb{R})$ with all *translations and rotations*, then $L^2(\mathbb{R})$ has *no* nontrivial, closed, invariant subspaces. That is, we will show that $L^2(\mathbb{R})$ is *irreducible* with respect to the combined action of translations and rotations.

By *rotations*, in this context, we mean all operators that act on functions $f \in L^2(\mathbb{R})$ by multiplication M_{χ_α} by a character $\chi_\alpha(x) = e^{2\pi i \alpha x}$. Thus

$$(M_{\chi_\alpha} f)(x) = \chi_\alpha(x) f(x)$$

for all $x \in \mathbb{R}$. A rotation rotates the *values* of f in the complex plane. The reader should note easily that M_χ does map $L^2(\mathbb{R})$ into itself. Thus we will prove here that if H is a closed, nontrivial subspace of $L^2(\mathbb{R})$ that is invariant under all translations and all multiplications by characters, then $H = L^2(\mathbb{R})$. After this work is done, we will explain the connection of this theorem with the Heisenberg group, the Heisenberg Uncertainty Principle, the Schrödinger model of the position and momentum operators in quantum mechanics, and a theorem of Stone and von Neumann.

Theorem 10.5.1 *Let H be any closed, nontrivial subspace of $L^2(\mathbb{R})$ that is invariant under the actions of all the multiplications $M_{\chi_\alpha}(f)(x) = \chi_\alpha(x)f(x)$ and invariant under all the translations $T_x(f)(t) = f(t+x)$. Then $H = L^2(\mathbb{R})$.*

Proof: By Theorem 10.4.1, there exists a measurable set $\widehat{H}$, called the *spectrum* of H, such that

$$H = \left\{ f \in L^2(\mathbb{R}) \,\middle|\, \widehat{f}\,\big|_{\widehat{H}^c} = 0 \text{ a.e.} \right\}.$$

We need to prove that $\widehat{H}^c$ is a Lebesgue null set. Suppose that $l\left(\widehat{H}^c\right) > 0$. We will deduce a contradiction.

Since H is nontrivial, there exists a function $f \neq 0 \in L^2(\mathbb{R})$ for which $f \in H$. Thus the set $S_f = \left\{ \alpha \in \mathbb{R} \,\middle|\, \widehat{f}(\alpha) \neq 0 \right\}$ has strictly positive Lebesgue measure. We know from any one of the Exercises 3.26, 6.11, or 7.21 [125] that there exists $\beta \in \mathbb{R}$ such that

$$l\left((\beta + S_f) \bigcap \widehat{H}^c \right) > 0.$$

We claim that

$$(M_{\chi_\beta} f)\widehat{}(\alpha) = \widehat{f}(\alpha - \beta),$$

for almost all α. This would imply that

$$S_{(M_{\chi_\beta} f)} = \beta + S_f.$$

For functions in $L^1(\mathbb{R})$ this follows immediately from the definition of the Fourier transform. If $f \in L^2(\mathbb{R})$ we can define $f_n = f\big|_{[-n,n]}$, so that f_n lies in $L^1(\mathbb{R})$ for each n and $f_n \to f$ in the L^2-norm. By passing to a suitable subsequence f_{n_i} of the sequence f_n, we can be assured that $f_{n_i} \to f$ almost everywhere as well, and this proves the claim.

The proof of the theorem is complete, since $M_{\chi_\beta} f \in H$ and because $\widehat{H}^c$ is disjoint from $\widehat{H}$. ∎

---

[125]The reader may enjoy noting how any one of the measure-theoretic exercises cited here can play the crucial role in the proof of this theorem.

10.5.1 Position and Momentum Operators

It is natural to consider the action of $\mathbb{R}$ by translation upon $L^2(\mathbb{R})$, but the reader may wonder why we consider here the combined actions of translation and rotation on $L^2(\mathbb{R})$. In order to address this question, we present a brief description of the quantum mechanical formalism for the position and momentum operators of one isolated quantum (particle) having only one degree of freedom, meaning that it is able to move only along the real line, $\mathbb{R}$. Our discussion is only a sketch, and we will not concern ourselves with physical constants, however important, treating them as though they were 1 wherever this is convenient. Our main purpose is to explain how the combined action of translations and rotations stems from the action of a *nonabelian group*, called the *Heisenberg group*, on $L^2(\mathbb{R})$ and what this action has to do with physics.

The *state* of the particle is interpreted as being a complex-valued function ϕ in $L^2(\mathbb{R})$, having the additional property that $\|\phi\|_2 = 1$. This makes $|\phi|^2$ into a probability density function, meaning that the probability that the position x of the quantum is between the coordinates a and b is given by

$$\mathcal{P}(a \leqslant x \leqslant b) = \int_a^b |\phi(x)|^2 \, dx.$$

The *expected value* of the position is given by

$$\langle x\phi, \phi \rangle = \int_{\mathbb{R}} x|\phi(x)|^2 \, dx.$$

It is thus natural to define the *position operator* P by

$$(P\phi)(x) = x\phi(x),$$

so that the expectation of the position is $\langle P\phi, \phi \rangle$, and this scalar product is a real number, although the function ϕ is complex-valued. Note that the domain of P is

$$D_P = \{\phi \in L^2(\mathbb{R}) \,|\, x\phi \in L^2(\mathbb{R})\}$$

and that this is a dense, but not closed, subspace of $L^2(\mathbb{R})$.

Conceptually, the quantum can be regarded as *being* the state function ϕ, and it can be interpreted as being present at all locations in the support of ϕ. Probability enters the picture as soon as one makes a macroscopic observation or measurement to try to detect the presence of the quantum. As evidence for this abstract notion of the position of a particle, physicists cite an experiment in which a single quantum is released on one side of a barrier that has two parallel slits cut into it. If one places detectors at the two slits, it will turn out that the quantum passes through either one slit or the other—not through both. The probability of the quantum being located at either slit is governed by the probability distribution ϕ. However, if no detectors are placed at the slits and a detection screen is placed opposite the wall that separates the quantum from the detector, then it turns out that a diffraction pattern appears on the screen. The pattern that is produced shows that the quantum has passed wave-like through both slits, creating a diffraction pattern on the detection screen by interference with itself.

The Fourier transform $\hat{\phi}$ permits one to express ϕ as an integral of characters χ_ν. Each index ν corresponds to a pure frequency having an energy level $E = h\nu$, h being Planck's constant. The index ν is taken also as corresponding to momentum. We note that $\left\| \hat{\phi} \right\|_2 = 1$, thanks to the Plancherel identity. Thus $\left| \hat{\phi} \right|^2$ is a probability density function. The probability that the momentum M is between a and b is given by

$$ P(a \leqslant M \leqslant b) = \int_a^b \nu \left| \hat{\phi}(\nu) \right|^2 d\nu. $$

The expected value of the momentum is given by

$$ \left\langle \nu \hat{\phi}, \hat{\phi} \right\rangle = \int_{\mathbb{R}} \nu \left| \hat{\phi}(\nu) \right|^2 d\nu. $$

However,

$$ \nu \hat{\phi} = \frac{1}{2\pi i} \widehat{\left(\frac{d}{dx} \phi \right)}(\nu) = \frac{-i}{2\pi} \widehat{\phi'}(\nu). $$

Hence we define the *momentum operator* Q by

$$ Q\phi = \frac{-i}{2\pi} \phi'. $$

and the domain D_Q of Q is the set of those square integrable functions ϕ such that the derivative ϕ' exists and is square integrable. This domain is a dense, but not closed, subspace of $L^2(\mathbb{R})$. The reader will prove in Exercise 10.19 that

$$ PQ - QP = \frac{i}{2\pi} I, \tag{10.14} $$

where I is the identity operator, restricted to the domain

$$ D = Q^{-1}(D_P) \cap P^{-1}(D_Q). $$

According to the Heisenberg Uncertainty Principle, the failure of the operators P and Q to commute with one another means that the result of the combination of position and momentum operators is dependent upon the *order* in which the two operators are applied. Physically, each measurement (of position or momentum) alters the subsequent measurement of the other. Thus it matters which operation is performed first and which second.

10.5.2   The Heisenberg Group

In our brief survey of the quantum mechanical formalism for position and momentum, we have derived Equation (10.14) from the definitions given here for the position and momentum operators, P and Q, respectively. More generally, Equation (10.14) is taken as fundamental, and the formulas given here for P and Q are Schrödinger's model (concrete realization, or example) of operators satisfying the Heisenberg Commutation Relation. A fundamental mathematical question that arises is whether or not Schrödinger's model is unique in some suitable sense as an operator solution to Equation (10.14).

The Heisenberg group, $\mathbb{H}$, is the group of all real matrices of the form

$$(x, y, z) = \begin{pmatrix} 1 & x & z \\ 0 & 1 & y \\ 0 & 0 & 1 \end{pmatrix},$$

with the operation being ordinary matrix multiplication. Since these upper-triangular matrices depend only upon the three real parameters x, y, and z, we denote the matrix by the more efficient symbol (x, y, z), understanding this to represent the full 3×3 matrix. The reader should check easily that the multiplication in the Heisenberg group is given by

$$(x, y, z)\,(x', y', z') = (x + x', y + y', z + z' + xy')$$

and that this multiplication is nonabelian. (See Exercises 10.20 and 10.21.)

For each $(x, y, z) \in \mathbb{H}$, we define the following action on $L^2(\mathbb{R})$:

$$\pi_{(x,y,z)}\phi(t) = \chi_1(z)\chi_y(t)\,(T_x\phi)\,(t) \text{ or} \tag{10.15}$$

$$\pi_{(x,y,z)}\phi = \chi_1(z)M_{\chi_y}T_x\phi,$$

where T_x denotes translation by x and M_{χ_y} denotes multiplication by the character χ_y. Of course, L^2-functions are defined only almost everywhere, so the preceding equation should be understood as applying pointwise almost everywhere. The reader will show in Exercise 10.22 that π is a continuous homomorphism of the group $\mathbb{H}$ into the group of norm-preserving Hilbert space automorphisms of $L^2(\mathbb{R})$. We see at once that the operators $\pi_{(x,0,0)}$ for all $x \in \mathbb{R}$ include all translation operators on $L^2(\mathbb{R})$. The operators $\pi_{(0,y,0)}$ provide all the rotation operators. Theorem 10.5.1 tells us that $L^2(\mathbb{R})$ has no nontrivial closed, proper invariant subspaces under the action of π. The representation π is said to be *irreducible* because of the absence of nontrivial, closed, π-invariant subspaces. It is called *unitary* because π acts in such a way as to preserve the Hermitian scalar product of $L^2(\mathbb{R})$.

It is a simple calculation to show that

$$P = \frac{1}{2\pi i}\frac{\partial}{\partial y}\Big|_{y=0}\pi_{(0,y,0)} \text{ and} \tag{10.16}$$

$$Q = \frac{1}{2\pi i}\frac{\partial}{\partial x}\Big|_{x=0}\pi_{(x,0,0)}.$$

Here it must be understood again that P and Q are defined only on that dense subspace of $L^2(\mathbb{R})$ consisting of functions that have images under P and Q that remain in $L^2(\mathbb{R})$. We see that the position and momentum operators of Schrödinger arise naturally by differentiation of the representation π of the Heisenberg group, $\mathbb{H}$. In this context, the uniqueness property of the Schrödinger model for the solution operators P and Q of the Heisenberg Commutation Relation is established by a famous theorem of Stone and von Neumann. This theorem asserts that the representation π is determined uniquely up to isomorphism of Hilbert space by its restriction to the center

$$\mathcal{Z}(\mathbb{H}) = \{(0, 0, z) \mid z \in \mathbb{R}\}.$$

In other words, any other representation σ, having the same restriction to the center, must have the property that there is an isomorphism τ of Hilbert space such that

$$\tau \circ \pi_{(x,y,z)} \equiv \sigma_{(x,y,z)} \circ \tau$$

for all $(x, y, z) \in \mathbb{H}$.

The reader who would like to study carefully the ideas sketched in the present section is referred to [19] and [4].

EXERCISES

10.19 Prove Equation (10.14) by showing that

$$(PQ - QP)\phi = \frac{i}{2\pi}\phi,$$

the Heisenberg *Commutation Relation* for the position and momentum operators in quantum mechanics.

10.20 Show that the Heisenberg group is a nonabelian group, closed under both multiplication and inversion.

10.21 Show that Lebesgue measure on $\mathbb{R}^3$ is invariant under the operation of right translation by an arbitrary element of the Heisenberg group. That is, define for each point (a, b, c) in $\mathbb{R}^3$, the mapping $T_{(a,b,c)} : \mathbb{R}^3 \to \mathbb{R}^3$ by

$$T_{(a,b,c)}(x, y, z) = (x + a, y + b, z + c + xb).$$

Show that this map preserves both Lebesgue measurability and Lebesgue measure. Show that Lebesgue measure is invariant under left translation as well.

10.22 Show that the action π defined by Equation (10.15) is a homomorphism of the multiplicative group $\mathbb{H}$ into the group of all norm-preserving Hilbert space automorphisms of the space $L^2(\mathbb{R})$. In particular, show that

$$\pi_{(x,y,z)}\pi_{(x',y',z')} = \pi_{(x+x', y+y', z+z'+xy')},$$

and show that π is continuous in the sense that $\left\|\pi_{(x,y,z)}\phi - \phi\right\|_2 \to 0$ for each $\phi \in L^2(\mathbb{R})$, as $(x, y, z) \to (0, 0, 0)$ in the sense of convergence in $\mathbb{R}^3$.

10.23 Prove Equations (10.16) by direct calculation.

APPENDIX:

THE BANACH-TARSKI THEOREM

A.1 THE LIMITS TO COUNTABLE ADDITIVITY

In Section 1.1, we considered the pivotal role of the discovery of incommensurable line segments in the development of Euclidean geometry. We identified commensurability problems as belonging to the early development of measure theory. In the pages that followed, we have learned much about Lebesgue's theory of measure and integration. We have seen how the Lebesgue theory yields complete normed linear spaces such as $L^p(X, \mathfrak{A}, \mu)$. We have learned about the dual spaces of the latter spaces, and about the dual of the space $\mathcal{C}(X)$, if X is a compact Hausdorff space. It would be difficult to overstate the importance of Lebesgue measure and integration throughout modern pure and applied analysis.

The reader has seen that much effort must be made to ensure that we deal only with measurable sets and measurable functions in Lebesgue's theory. We showed in Example 3.1, using the Axiom of Choice, that no translation-invariant, countably additive measure μ can be defined on *all* the subsets $E \subseteq \mathbb{R}$ if

$$0 < \mu[0, 1) < \infty.$$

Measure and Integration: A Concise Introduction to Real Analysis. By Leonard F. Richardson
Copyright © 2009 John Wiley & Sons, Inc.

In this appendix, we will discuss a theorem that is as challenging for modern mathematicians as incommensurable line segments were for their ancient Greek forerunners. It is a theorem of Banach and Tarski, and it is often called the *Banach-Tarski Paradox*. It is not a paradox in fact, but a theorem. This theorem is described as a *paradox* because of the extraordinarily counterintuitive nature of its conclusions.

We begin with a definition.

Definition A.1.1 Two sets A and B in a metric space (X, ρ) are said to be *congruent by finite decomposition* provided that there is a natural number n such that it is possible to decompose A and B into disjoint unions

$$A = \overset{\cdot}{\bigcup}_{1 \leqslant k \leqslant n} A_k \text{ and } B = \overset{\cdot}{\bigcup}_{1 \leqslant k \leqslant n} B_k$$

in such a way that A_k and B_k are *congruent* for each $k \leqslant n$. This is denoted by

$$A \overset{f}{\cong} B.$$

Congruence means that there exists a *bijection* $T_k : A_k \to B_k$ such that

$$\rho(x, y) = \rho(T_k x, T_k y)$$

for all x and y in A_k. Such a mapping T_k is also called a *bijective isometry*. We denote congruence as

$$A_k \cong B_k.$$

We call A and B *congruent by countable decomposition* provided that the decomposition of A and B into mutually congruent pieces can be accomplished by using countably many pieces. This is denoted by

$$A \overset{c}{\cong} B.$$

One may prove readily as an exercise that a linear isometry of Euclidean space must preserve the measure of any Lebesgue measurable set.

The reader will note that Steinhaus's[1] theorem (7.4.3) shows that two sets of the *same* Lebesgue measure in the real line must be congruent by countable decomposition. The Banach-Tarski theorem is much more startling.

Theorem A.1.1 (Banach-Tarski) *Let A and B be two subsets of $\mathbb{R}^n$, each having nonempty interior, and with $n \geqslant 3$. Then $A \overset{f}{\cong} B$. (For example, any two spherical balls are congruent by finite decomposition regardless of the difference in their radii.) For dimensions 1 and 2, $A \overset{c}{\cong} B$.*

We remark that this theorem implies that for $\mathbb{R}^n$, with $n \geqslant 3$, there cannot exist even a *finitely* additive measure defined on all subsets that is invariant under Euclidean

---

[1] As an historical sidelight, we remark that Steinhaus played a pivotal role in Banach's decision to become a professional mathematician. Further information is available in [17].

motions. For $n = 1$ or $n = 2$, there is no countably additive measure that is defined on all subsets that is invariant under Euclidean motions. That is, in each of these two sets of circumstances, nonmeasurable sets must exist.

The proof given by Banach and Tarski can be found in their original paper [2]. There is also a modern treatise on the subject [25]. The reasoning comes primarily from abstract set theory and from the study of groups of linear transformations acting on vector spaces.

We give a simple example below that illustrates the Banach-Tarski theorem.

■ EXAMPLE A.1

We will show that there exists a subset $S \subset [0, 2)$, in the real line, for which $S \overset{c}{\cong} \mathbb{R}$. That is, $S \subset [0, 2)$ will be congruent by countable decomposition to the entire real line. The congruence mappings will be translations.

The example begins with the proof in Example 3.1 that there exists a non-measurable subset of the line. There we defined an equivalence relation:

$$x \sim y \Leftrightarrow x - y \in \mathbb{Q}.$$

Note that each real number is equivalent modulo rational translation to numbers in the interval $[0, 1)$. We use the Axiom of Choice to select an uncountable set C in $[0, 1)$ having the property that C consists of one element from each equivalence class in $\mathbb{R}/\sim$. The set C is called a *cross section* of $\mathbb{R}/\sim$. If we were still in Example 3.1, we would proceed to explain why the set C must be nonmeasurable. Instead, the example takes a surprising turn. [2]

Let $U = \mathbb{Q} \cap [0, 1]$, and let $C_q = C + q$. Define

$$S = \dot{\bigcup}_{q \in U} (C_q),$$

the *disjoint* union of the countably many translates of C by $q \in U$. Thus $S \subseteq [0, 2)$. Let

$$\tau : U \to \mathbb{Q}$$

be a *bijection* between the two countable sets, U and $\mathbb{Q}$. Let

$$x_q = \tau(q) - q$$

for each $q \in U$. Observe that

$$\mathbb{R} = \dot{\bigcup}_{q \in U} (C_q + x_q),$$

as claimed.

---

[2] The author learned the following example from the website of Professor Terence Tao at UCLA. This surprise ending for the famous example of a nonmeasurable set deserves to be better known, because it is so simple, compelling, and delightful.

This example shows with convincing simplicity an instance of a congruence, by countable decomposition, between two seemingly *incongruous* sets. It provides a compelling explanation of why it is necessary to check for measurability in analysis. It gives also a startling sense of the geometrical possibilities of nonmeasurable sets.

The theorem of Banach and Tarski demonstrates that the tools of Lebesgue measure and integration, with all their strength, cannot measure all sets, even with *finite* additivity. More striking is the capacity of nonmeasurable sets to be reconfigured in astounding ways, with geometrical perfection, losing nary a point.

Thanks to Lebesgue and many other mathematicians, the understanding of measure and integration has been advanced in many ways that are invaluable to analysis. Yet the mysteries that stand are more profound than those that came before. This is a challenge and an invitation to the student to engage in the search to expand human knowledge. It affirms for us the grandeur of truth, the finiteness of ourselves, and the good fortune to be allowed a glimpse.

REFERENCES

1. S. Banach, Sur l'équation fonctionnelle $f(x + y) = f(x) + f(y)$. *Fundamenta Mathematicae*, Tom 1, Warszawa, 1920.[1]

2. S. Banach and A. Tarski, Sur la décomposition des ensembles en parties respectivement congruentes. *Fundamenta Mathematicae*, Tom 6, Warszawa, 1924.[1]

3. Lennart Carleson, On convergence and growth of partial sumas of Fourier series. *Acta Mathematica*, Vol. 116, 1966, pp. 135–157.

4. L. Corwin and F. P. Greenleaf, *Representations of Nilpotent Lie Groups and Their Applications*. Cambridge University Press, Cambridge, 1990.

5. H. Dym and H. P. McKean, *Fourier Series and Integrals*. Academic Press, New York, 1972.

6. J. B. J. Fourier, *Théorie Analytique de la Chaleur*. Firmin Didot, Paris, 1822.

7. B. Gelbaum and J. Olmsted, *Counterexamples in Analysis*. Holden-Day, San Francisco, 1964.

8. C. Goffman, *Real Functions*. Rinehart and Company, New York, 1953.

9. C. Goffman and G. Pedrick, *First Course in Functional Analysis*. Prentice-Hall, Englewood Cliffs, NJ, 1965.

---

[1]This historic paper is available online from the Institute for Computational Mathematics at the University of Warsaw in Poland at http://matwbn.icm.edu.pl/.

10. P. Halmos, *Measure Theory*. D. van Nostrand Company, New York, 1950.

11. G. Hamel, Eine basis aller zahlen und die unstetigen lösungen des funktionalgleichung $f(x + y) = f(x) + f(y)$. *Mathematische Annalen*, Vol. 60, 1905, pp. 459–462.

12. Kenneth Hoffman and Ray Kunze, *Linear Algebra*. Prentice-Hall, Englewood Cliffs, NJ, 1971.

13. P. Jordan and J. v. Neumann, On inner products in linear, metric spaces. *Annals of Mathematics*, Vol. 36, No. 3, 1935, pp. 719–723.

14. S. Kakutani, Concrete representation of abstract (M)-spaces (A characterization of the space of continuous functions). *Annals of Mathematics*, 2nd Ser., Vol. 42, No. 4, pp. 1941, 994–1024.

15. S. Kakutani, A proof of the uniqueness of Haar's measure. *Annals of Mathematics*, Vol. 49, 1948, pp. 225–226.

16. E. Kamke, *Theory of Sets*, translated by F. Bagemihl. Dover Publications, New York, 1950.

17. *The MacTutor History of Mathematics*, Archive of the University of St. Andrews, Fife, Scotland.[2]

18. L. Nachbin, *The Haar Integral*. D. van Nostrand Company, New York, 1965.

19. L. Pukanszky, *Leçons sur les Représentations des Groupes*. Dunod, Paris, 1967.

20. L. Richardson, *Advanced Calculus: An Introduction to Linear Analysis*. John Wiley & Sons, 2008.

21. F. Riesz and B. Sz.-Nagy, *Functional Analysis*. Frederick Ungar Publishing Company, New York, 1955.

22. S. Saks, *Theory of the Integral*, 2nd ed., translated by L. C. Young. Hafner Publishing Company, New York. (First ed. Warsaw, 1937.[3])

23. S. Saks, Integration in abstract metric spaces. *Duke Mathematics Journal*, Vol. 4, 1938, pp. 408–411.

24. G. E. Shilov and B. L. Gurevich, *Integral, Measure and Derivative: A Unified Approach*, translated by R. A. Silverman. Prentice-Hall, Engleewood Cliffs, NJ, 1966.

25. S. Wagon, *The Banach-Tarski Paradox*. Cambridge University Press, Cambridge, 1986.

---

[2]This archive is at http://www-history.mcs.st-andrews.ac.uk/history/index.html.
[3]This historic book is available online from the Institute for Computational Mathematics at the University of Warsaw in Poland at http://matwbn.icm.edu.pl/.

INDEX

Measure and Integration: A Concise Introduction to Real Analysis. By Leonard F. Richardson   **231**
Copyright © 2009 John Wiley & Sons, Inc.

Printed and bound by CPI Group (UK) Ltd, Croydon, CR0 4YY

16/04/2025

14658368-0001